SAMPLING –
MICROBIOLOGICAL MONITORING
OF ENVIRONMENTS

THE SOCIETY FOR APPLIED BACTERIOLOGY
TECHNICAL SERIES NO. 7

SAMPLING – MICROBIOLOGICAL MONITORING OF ENVIRONMENTS

Edited by

R. G. BOARD

School of Biological Sciences, University of Bath, Claverton Down, Bath, England

AND

D. W. LOVELOCK

H. J. Heinz Co. Ltd., Hayes Park, Hayes, Middlesex, England

1973

ACADEMIC PRESS · LONDON · NEW YORK

ACADEMIC PRESS INC. (LONDON) LTD
24/28 OVAL ROAD
LONDON, NW1

U.S. Edition published by
ACADEMIC PRESS INC.
111 FIFTH AVENUE,
NEW YORK, NEW YORK 10003

Library of Congress Catalog Card No : 73-5015
ISBN : 0–12–108250–4

Printed in Great Britain by
Cox & Wyman Ltd., Fakenham, Norfolk, England

Contributors

M. C. ALLWOOD, *Department of Pharmacy, University of Manchester, Manchester, England*

ELLA M. BARNES, *Food Research Institute, Colney Lane, Norwich NOR 70F, England*

W. CLIFFORD, *Food Hygiene Laboratory, Central Public Health Laboratory, Colindale Avenue, London NW9 5HT, England*

C. B. COLE, *National Institute for Research in Dairying, University of Reading, Shinfield, Reading RG2 9AT, England*

VERA G. COLLINS, *Freshwater Biological Association, The Ferry House, Ambleside, Westmorland, England*

G. L. CURE, *Department of Microbiology, The University, Reading RG1 5AQ, England*

J. F. DARBYSHIRE, *Department of Microbiology, The Macaulay Institute for Soil Research, Craigiebuckler, Aberdeen AB9 2QJ, Scotland*

R. R. DAVENPORT, *University of Bristol, Department of Agriculture and Horticulture, Research Station, Long Ashton, Bristol, England*

G. A. GARDNER, *Ulster Curers' Association, 2 Greenwood Avenue, Belfast BT4 3JL, Northern Ireland*

A. C. GHOSH, *Food Hygiene Laboratory, Central Public Health Laboratory, Colindale Avenue, London NW9 5HT, England*

R. J. GILBERT, *Food Hygiene Laboratory, Central Public Health Laboratory, Colindale Avenue, London NW9 5HT, England*

J. M. GRAINGER, *Department of Microbiology, The University, Reading RG1 5AQ, England*

T. R. G. GRAY, *Hartley Botanical Laboratories, P.O. Box 147, University of Liverpool, Liverpool L69 3BX, England*

MARGARET S. HENDRIE, *Torry Research Station, 135 Abbey Road, Aberdeen AB9 8DG, Scotland*

BETTY C. HOBBS, *Food Hygiene Laboratory, Central Public Health Laboratory, Colindale Avenue, London NW9 5HT, England*

A. J. HOLDING, *Department of Microbiology, School of Agriculture, University of Edinburgh, Edinburgh EH9 3JG, Scotland*

W. R. HUDSON, *Meat Research Institute, Langford, Bristol BS18 7DY, England*

C. S. Impey, *Food Research Institute, Colney Lane, Norwich NOR 70F, England*

G. C. Ingram, *Meat Research Institute, Langford, Bristol BS18 7DY, England*

J. G. Jones, *Freshwater Biological Association, The Ferry House, Ambleside, Westmorland, England*

R. M. Keddie, *Department of Microbiology, The University, Reading RG1 5AQ, England*

Margaret Kendall, *Food Hygiene Laboratory, Central Public Health Laboratory, Colindale Avenue, London NW9 5HT, England*

A. G. Kitchell, *Meat Research Institute, Langford, Bristol BS18 7DY, England*

B. A. Law, *National Institute for Research in Dairying, University of Reading, Shinfield, Reading RG2 9AT, England*

L. A. Mabbitt, *National Institute for Research in Dairying, University of Reading, Shinfield, Reading RG2 9AT, England*

R. T. Parry, *Bernard Matthews Ltd., Great Witchingham Hall, Norwich NOR 65X, England*

Muriel E. Rhodes, *Department of Botany and Microbiology, University College of Wales, Aberystwyth, Wales*

Diane Roberts, *Food Hygiene Laboratory, Central Public Health Laboratory, Colindale Avenue, London NW9 5HT, England*

A. N. Sharpe, *Unilever Research Laboratory Colworth/Welwyn, Colworth House, Sharnbrook, Bedford, England*

M. Elisabeth Sharpe, *National Institute for Research in Dairying, University of Reading, Shinfield, Reading RG2 9AT, England*

J. M. Shewan, *Torry Research Station, 135 Abbey Road, Aberdeen AB9 8DG, Scotland*

A. H. Varnam, *Department of Microbiology, The University, Reading RG1 5AQ, England*

Antonnette A. Wieneke, *Food Hygiene Laboratory, Central Public Health Laboratory, Colindale Avenue, London NW9 5HT, England*

S. T. Williams, *Hartley Botanical Laboratories, P.O. Box 147, University of Liverpool, Liverpool L69 3BX, England*

D. D. Wynn-Williams, *Department of Botany and Microbiology, University College of Wales, Aberystwyth, Wales*

Preface

THIS volume includes contributions to the Autumn Demonstration Meeting of the Society for Applied Bacteriology, held on 27 October 1971 at the Unilever Research Laboratory, Isleworth, London. It is Number 7 in the Technical Series and it marks the continuation of the Society's policy of providing experts in a field with the opportunity of demonstrating methods and techniques to Members and guests of the Society. As in previous years, the demonstrators have described their methods in this book which is intended primarily as a ready reference source at the laboratory bench. Although the demonstrators came only from laboratories in the United Kingdom, we feel that the methods and techniques which they describe in this volume will prove useful to their colleagues the world over. We wish to thank them for the great effort which they took in the preparation of the demonstrations and their chapters in this book.

Our thanks go also to Mr. R. Woodroffe and his colleagues at the Unilever Research Laboratories, Isleworth, for their help with the laboratory arrangements for the demonstrations.

January, 1973

R. G. BOARD
D. W. LOVELOCK

Contents

Microflora of Cheddar Cheese and Some of the Metabolic Products

B. A. Law, M. Elisabeth Sharpe, L. A. Mabbitt and C. B. Cole

National Institute for Research in Dairying, University of Reading, Shinfield, Reading RG2 9AT, England

The microflora of newly formed cheese curd is mainly composed of the starter organisms deliberately added to develop the lactic acid necessary for curd formation. The starter culture may include *Streptococcus lactis*, *S. cremoris*, *S. diacetilactis*, and leuconostocs (Reiter and Møller Madsen, 1963). In addition, adventitious microorganisms are also present in the curd; these comprise organisms originally present in the raw milk and which have survived the heat treatment given to all commercially made cheese milk in the United Kingdom, together with contaminants from the factory equipment, the air, and from dairy personnel. All these organisms, which may include many different groups, may be present in the young cheese. To assess their possible contribution to Cheddar cheese ripening, it is necessary to follow the survival or multiplication of all groups during cheese maturation. For other purposes it may only be required to follow the survival of the starter streptococci or of the different component species in a commercial mixed starter, or to detect potentially toxin forming organisms such as coagulase positive staphylococci, or to detect possible spoilage organisms such as coliforms.

Cheese Sampling

Equipment

Scissors and scalpel are immersed in ethanol and flamed just before use. Sterile steel corers, and plungers to fit inside the corers, are wrapped separately in greaseproof paper and autoclaved. When a series is to be taken from a cheese at different stages of maturation, a stainless steel template of the cheese surface with a pattern of equidistant holes is useful to obtain equidistant samples (Fig. 1). Paraffin wax (Hansen's cheese wax, tropical grade) is sterilized by autoclaving in 15 ml amounts in plugged tubes, with a wick of string down the centre of the wax.

Sampling variation

Because the starter organisms are added in large numbers and are well stirred with the cheese milk they are distributed homogeneously throughout the cheese. In comparison, the adventitious bacteria gain access before

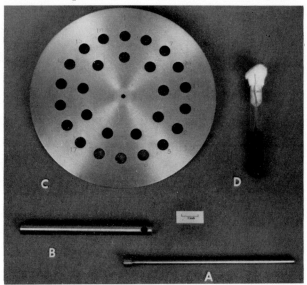

FIG. 1. Equipment used for aseptic sampling of cheese. A, B, plunger and corer; C, stainless steel template with a pattern of equidistant holes, and D, plugged tube containing Hansen's cheese wax. For additional details see text.

TABLE 1. Relative concentrations* of volatile substances at different points in a Cheddar cheese

Volatile substance		Distance from cheese surface (mm)				
		25	35	45	55	65
Cheese A	Butan-2-one	6·25†	2·90	1·60	1·0	0·95
	Ethanol	3·40	3·10	3·40	3·60	4·25
Cheese B	Butan-2-one	6·25	2·95	1·5	0·95	0·85
	Ethanol	3·45	3·60	4·1	3·62	4·0

* The cheese was bored with a cheese corer (diam 12 mm) and the resultant cheese bore was cut transversely into 5 mm thick slices. Each slice was separately equilibrated at 30° in a closed bottle (5 ml) and 1 ml of the "head space" analysed by gas liquid chromatography.

† Concentrations are expressed as ratios of the concentration of the component mentioned to that of acetone.

or at any time during the cheesemaking process, in some cases after the curd has set, and are consequently heterogeneously distributed (Naylor and Sharpe, 1958a, b; Dean, Berridge and Mabbitt, 1959). The different relative amounts of volatile bacterial products found at different points in the cheese (Table 1) may be partially caused by this non-uniform distribution. It is therefore important, particularly when sampling for the non-starter microflora, to ensure that the sample is representative. This is achieved by grating and mixing three samples and sub-sampling the bulk.

Sampling

A single sample consists of three 6 in. long, $\frac{1}{2}$ in. diam bores taken from different parts of the cheese, either equidistant on the top, or from top and sides. Before removing each bore, the exterior of the cheese is swabbed with ethanol, an area 1 in. in diam of the outer plastic or wax coating cut out with scissors and $\frac{1}{2}$ in. deep layer of cheese removed with a scalpel. A bore of cheese is then taken with the corer, through the newly exposed surface and each bore extruded by the plunger into a sterile Petri dish. The top $\frac{1}{2}$ in. of the bore is cut off and discarded. The hole in the cheese is filled with sterile, partially melted wax, which is handled by its wick (Fig. 1), and topped up with melted wax. The wax reseals the cheese against outside contamination and excessive aeration and provides a solid filling to prevent texture deterioration. The use of only partially melted wax prevents wax diffusing too far into the cheese cracks.

Processing of sample

Equal portions of each bore are grated together aseptically in a sterilized (by autoclaving) Mouli (Moulinex, London) grater, and 10 g of the well mixed grated cheese is homogenized in an overhead-drive blender for 3 min with 90 ml of 1% (w/v) sodium citrate, previously warmed to 50°. The homogenate is then allowed to stand for 10 min to allow the foam to settle (Naylor and Sharpe, 1958a).

Plating of sample

Suitable dilutions of the homogenate in quarter-strength Ringer's solution are plated on the different media required for the particular microflora being examined. For acetate agar and yeast glucose agar, deep plates are poured. For the other media surface plating is used. Acetate agar plates are incubated anaerobically in an atmosphere of H_2 (90%)/CO_2 (10%), other media being incubated aerobically. All plates are incubated at 30°

for 3 days, except tetrazolium thallous acetate plates which are incubated at 45° for 2 days.

Enumeration and Isolation of Microorganisms

Media

Apart from the general purpose medium described below, all the others are selective or indicator media.

General purpose medium

Yeast glucose agar—YGA—(Naylor and Sharpe, 1958a) is used for all groups of organisms, and also for starter streptococci in samples taken during the early stages of cheese maturation, when they predominate. Colonies of starter streptococci are picked into Yeast Glucose broth (YGB) and purified on YGA. Other organisms are not usually picked from this medium.

Components of starter microflora

Citrate fermenting (aroma producing) *S. diacetilactis* and leuconostocs and non-fermenting, *S. lactis* and *S. cremoris* strains of starter organisms can be distinguished on the indicator medium described by Galesloot, Hassing and Stadhouders (1961). *S. lactis* and *S. cremoris* have been distinguished on an arginine indicator medium (Reddy, Vedamuthu, Washam and Reinbold, 1969).

Lactobacilli, pediococci, leuconostocs

Rogosa's acetate agar (AcA) of Rogosa, Mitchell and Wiseman (1951) is used. Colonies are picked into acetate broth and purified on MRS agar (de Man, Rogosa and Sharpe, 1960). On this medium there is no colonial differentiation between these three genera and initial microscopic identification is necessary.

Staphylococci and micrococci

Salt mannitol agar—SMA—(Chapman, 1945) which also differentiates mannitol positive from mannitol negative strains is used. Baird-Parker medium—BPA—(1962) is selective for coagulase positive staphylococci only. Colonies are picked from these media into nutrient broth and purified on nutrient agar.

Gram negative rods

Crystal violet (2 ppm) in nutrient agar (CV) is selective for these organisms Violet red bile agar—VRBA—(Oxoid) is selective for coliforms. Colonies are picked from these media into nutrient broth and purified on nutrient agar.

Group D streptococci

Thallous acetate tetrazolium glucose agar (TTA) of Barnes (1956) is used. Incubation at 45° inhibits the growth of starter streptococci which can also grow on this medium. Isolates are picked into YGB and purified on YGA.

Lipolytic bacteria

Numbers of tributyrin hydrolysing organisms are determined on tributyrin agar—TBA—(Franklin and Sharpe, 1963). Colonies showing clear zones of hydrolysis are picked into YGB and purified on YGA. The ability to hydrolyse tributyrin indicates esterase and/or lipase activity and is a screening test for lipolytic organisms. Active strains are then tested for their ability to hydrolyse butter fat by the method of Fryer, Lawrence and Reiter (1966).

Corynebacteria

These organisms occur as orange or yellow pigmented colonies on nutrient agar but only if the total non-lactic acid bacterial flora is small enough to allow them to be detected. Colonies are identified microscopically only and are not further examined.

Moulds

Mycological agar (Difco) is used, and also Yeast Maltose agar (YMA) with a composition similar to YGA except that maltose replaces glucose and the pH is adjusted to 5·5. Only the moulds, if any, are enumerated.

Identification of Isolates

The isolation of strains from specific selective media may be sufficient identification for many purposes. For further work the following schemes are used.

Lactobacilli, pediococci and leuconostocs

The methods of Perry and Sharpe (1960) and Sharpe, Fryer and Smith (1966) are used.

Staphylococci

Organisms in this group may be isolated as tributyrinolytic bacteria from TBA, as well as from SMA and BPA. The schemes of Shaw, Stitt and Cowan (1951) and of Baird-Parker (1966) are used.

Gram negative rods

These organisms are also found as tributyrin hydrolysers as well as isolates on CVA and VRBA. Further identification is by the scheme of Hendrie and Shewan (1966) for pseudomonads, of Carpenter, Lapage and Steel (1966) for Enterobacteriaceae, and of Thornley (1968) for achromobacters.

Group D streptococci

The classification described by Shattock (1962) and Sharpe *et al.* (1966) is used.

Viable Counts and Levels of Survival of Organisms in Cheddar Cheese

The following data obtained mainly in our own laboratories (Perry and Sharpe, 1960; Franklin and Sharpe, 1963; Reiter *et al.*, 1967) may be helpful to workers on cheese microflora.

Starter streptococci

The highest level of these organisms occurs in the newly formed curd and is $\sim 10^9$/g curd. During maturation the numbers decrease, the death rate depending on the strains present. *S. lactis* and *S. diacetilactis* survive for longer periods than *S. cremoris* (Perry, 1961).

Lactobacilli, pediococci and leuconostocs

These organisms may occur in only small numbers in the curd, depending on the level of contamination of the cheese milk, or the presence of heat resistant strains in the raw milk. They begin to multiply shortly after pressing, and usually reach their maximum numbers ($\sim 10^8$/g cheese) in 10–30 days; although they may be initially low in numbers in the curd (1–10/g curd) they may still multiply sufficiently to give a high final level. This high level may be maintained for at least 6 months and will then gradually decline. Pediococci occur less frequently than lactobacilli but multiply in a similar manner and may reach levels of $\sim 10^7$/g cheese (Fryer and Sharpe, 1966). Leuconostocs do not multiply in Cheddar cheese, but die out slowly.

Staphylococci and micrococci

These organisms may occur in numbers of 10^2–10^6/g in the cheese curd, depending on the heat treatment of the milk. They slowly decrease in numbers during ripening so that after 6 months maturation the level is 10^1–10^2/g cehese. Coagulase positive staphylococci, which are potential enterotoxin producers, are destroyed by the recommended heat treatment for cheese milk. Excessive multiplication after post pasteurization contamination of these organisms during cheesemaking can only occur if the starter produces insufficient acid, due to phage or antibiotics in the milk (Reiter, Fewins, Fryer and Sharpe, 1964; Sharpe, Fewins, Reiter and Cuthbert, 1965).

Gram negative rods

Psychrotrophs may be present in large numbers in bulk tank stored raw milks, but are killed by the heat treatment of the milk. However, some of their enzymes may survive and affect the cheese flavour (Sharpe, Fryer, Chapman and Reiter, 1970). Coliforms will multiply during the cheesemaking. All these organisms, if present, decline rapidly during the first few weeks of maturation.

Group D streptococci

These organisms may occur in numbers of 10^4-10^6/g curd, but in other cases may be present in very small numbers or be absent from the cheese curd. They decrease very irregularly, in some cheeses only surviving for a few weeks, in others showing little decrease in numbers after 3 months.

Corynebacteria

These occur in only very small numbers in the cheese, but often appear to survive for many months without increasing or decreasing.

Other organisms

Small numbers of spores of aerobic spore formers survive in the cheese for at least 6 months.

The number of non-starter bacteria occurring in the cheese is related to the history of the raw cheese milk and the amount of heat treatment it receives, and to the level of hygiene and cleanliness in the creamery. It is quite possible to have foci of infection of lactic acid bacteria such as lactobacilli and pediococci in the cheesemaking area which will lead to a high level of post pasteurization contamination of the cheese milk.

References

BAIRD-PARKER, A. C. (1962). An improved diagnostic and selective medium for isolating coagulase positive staphylococci. J. appl. Bact., 25, 12.

BAIRD-PARKER, A. C. (1966). Methods for classifying staphylococci and micrococci. Identification methods for microbiologists. Part A (B. M. Gibbs and F. A. Skinner, eds). London and New York: Academic Press.

BARNES, E. M. (1956). Methods for the isolation of faecal streptococci (Lancefield Group D) from bacon factories. J. appl. Bact., 19, 193.

CARPENTER, K. P., LAPAGE, S. P. & STEEL, K. J. (1966). Biochemical identification of Enterobacteriaceae. *Identification Methods for Microbiologists*. Part A (B. M. Gibbs and F. A. Skinner, eds). London and New York: Academic Press.

CHAPMAN, G. H. (1945). The significance of sodium chloride in studies of staphylococci. *J. Bact.*, **50**, 201.

DEAN, M. R., BERRIDGE, N. J. & MABBITT, L. A. (1959). Microscopical observations on Cheddar cheese and curd. *J. Dairy Res.*, **26**, 77.

FRANKLIN, J. G. & SHARPE, M. E. (1963). The incidence of bacteria in cheese milk and Cheddar cheese and their association with flavour. *J. Dairy Res.*, **30**, 87.

FRYER, T. F. & SHARPE, M. E. (1966). Pediococci in Cheddar cheese. *J. Dairy Res.*, **33**, 325.

FRYER, T. F., LAWRENCE, R. C. & REITER, B. (1966). Methods for isolation and enumeration of lipolytic organisms. *J. Dairy Sci.*, **50**, 477.

GALESLOOT, TH.E., HASSING, F. & STADHOUDERS, J. (1961). Agar medium for the isolation and enumeration of aromabacteria in starters. *Neth. Milk. Dairy J.*, **15**, 127.

HENDRIE, M. S. & SHEWAN, J. M. (1966). The identification of certain *Pseudomonas* species. *Identification Methods for Microbiologists*. Part A (B. M. Gibbs and F. A. Skinner, eds). London and New York: Academic Press.

MAN, J. C. DE, ROGOSA, M. & SHARPE, M. E. (1960). A medium for the cultivation of lactobacilli. *J. appl. Bact.*, **23**, 130.

NAYLOR, J. & SHARPE, M. E. (1958*a*). Lactobacilli in Cheddar cheese. I. The use of selective media for isolation and serological typing for identification. *J. Dairy Sci.*, **25**, 92.

NAYLOR, J. & SHARPE, M. E. (1958*b*). Lactobacilli in Cheddar cheese. II. Duplicate cheeses. *J. Dairy Res.*, **25**, 421.

PERRY, K. D. (1961). A comparison of the influence of *Streptococcus lactis* and *Str. cremoris* starters on the flavour of Cheddar cheese. *J. Dairy Res.*, **28**, 221.

PERRY, K. D. & SHARPE, M. E. (1960). Lactobacilli in raw milk and in Cheddar cheese. *J. Dairy Res.*, **27**, 267.

REDDY, M. S., VEDAMUTHU, E. R., WASHAM, C. J. & REINBOLD, G. W. (1969). Differential agar medium for separating *Streptococcus lactis* and *Streptococcus cremoris*. *Appl. Microbiol.*, **18**, 755.

REITER, B. & MØLLER MADSEN, A. (1963). Reviews of the progress of dairy science. Section B. Cheese and butter starters. *J. Dairy Res.*, **30**, 419.

REITER, B., FEWINS, B. G., FRYER, T. F. & SHARPE, M. E. (1964). Factors affecting the multiplication and survival of coagulase positive staphylococci in Cheddar cheese. *J. Dairy Res.*, **31**, 261.

REITER, B., FRYER, T. F., PICKERING, A., CHAPMAN, H. R., LAWRENCE, R. C. & SHARPE, M. E. (1967). The effect of the microbial flora on flavour and free fatty acid composition of Cheddar cheese. *J. Dairy Res.*, **34**, 257.

ROGOSA, M., MITCHELL, J. A. & WISEMAN, R. F. (1951). A selective medium for the isolation and enumeration of oral and fecal lactobacilli. *J. Bact.*, **62**, 132.

SHARPE, M. E., FRYER, T. F. & SMITH, D. G. (1966). Identification of the Lactic Acid Bacteria. *Identification Methods for Microbiologists*. Part A (B. M. Gibbs and F. A. Skinner, eds). London and New York: Academic Press.

SHARPE, M. E., FEWINS, B. G., REITER, B. & CUTHBERT, W. A. (1965). A survey of the incidence of coagulase-positive staphylococci in market milk and cheese in England and Wales. *J. Dairy Res.*, **32**, 187.

SHARPE, M. E., FRYER, T. F., CHAPMAN, H. R. & REITER, B. (1969–70). Report, National Institute for Research in Dairying, p. 148.

SHATTOCK, P. M. F. (1962). Enterococci. In symposium *Chemical and Biological Hazards in Foods*. Amos: Iowa State University Press.

SHAW, C., STITT, J. M. & COWAN, S. T. (1951). Staphylococci and their classification. *J. gen. Microbiol.*, **5**, 1016.

THORNLEY, M. J. (1968). Properties of *Acinetobacter* and related genera. *Identification Methods for Microbiologists*. Part B (B. M. Gibbs and D. A. Shapton, eds). London and New York: Academic Press.

The Microbiological Examination of Cured Meats

G. A. GARDNER

*Ulster Curers' Association, 2 Greenwood Avenue, Belfast BT4 3JL,
Northern Ireland*

AND

A. G. KITCHELL

Meat Research Institute, Langford, Bristol BS18 7DY, England

Our intention is not to review the innumerable published methods for the microbiological examination of cured meats, but to set out a basis upon which workers with interests in cured meats can develop their own schemes. The methods described are those which have been tested by us and shown to be of value; many of them form part of internationally recommended procedures. Discussion of many generally important aspects of sampling not peculiar to cured meats has been omitted because it will be found in other papers of this series.

It is essential to the understanding of the microbiology of cured meats to have knowledge of the chemical composition of the product in relation to the curing salts and, where feasible, simultaneous analyses for chloride (Schonherz, 1955), nitrate and nitrite (Follett and Ratcliff, 1963; Elliott and Porter, 1971), and water content (dry weight) should be made, and the pH should be determined.

Sampling

Cured meats are made up of skin, fat, muscle(s), connective tissue and, in some cases, even bone, and there is some indication that these can support different microfloras (Tofte Jespersen and Riemann, 1958; Ingram, 1960; Gardner and Patton, 1969). This can readily be demonstrated (J. A. Perigo, pers. comm.) by pressing a slice of bacon onto the surface of agar media in large dishes such as antibiotic assay plates. If selective media (e.g. a 7·5% w/v) NaCl medium for micrococci and an acetate medium for lactobacilli), are chosen it will be observed that the areas contacted by fat support mainly cocci and those contacted by lean, lactobacilli. In addition, meat is made up of surface and internal or deep tissue and it is known that different bacteria flourish in these two environments, the one aerobic, the other anaerobic. Thus if one is interested in a surface "sliming" condition, there is little point in including "deep tissue" with the sample, as this will result in a masking of the desired observation. It is necessary to distinguish such cases, to examine each independently, and assess the results in relation to the nature of the sample and the observed defect.

In general, it is desirable to use a destructive sampling technique, e.g. excision, if it is feasible. It is the method of choice for "spoiled" material or small goods, e.g. vacuum packed bacon. Non-destructive techniques are usually used for routine examination of sound material, particularly large joints or bacon sides, where destructive sampling would be too costly.

Non-destructive sampling

The use of cotton swabs for sampling surfaces has been practised for many years and it is still one of the most widely adopted non-destructive methods. The area of meat to be swabbed should be as large as is practical and representative of the bulk of the material. Where there are important differences between parts of the material (e.g. fat and lean, pleura and rind), swabs from each of these areas should be taken and either examined separately or bulked as a composite sample. The sampling of Wiltshire bacon sides provides a pertinent example. Five 10 cm^2 swabs, 3 from the rind and 2 from the inside surface (pleura), are bulked to give a 50 cm^2 composite sample for the side. Such a sampling system is useful when many sides of bacon have to be examined; when only a limited number is involved, it would be advantageous to increase the area swabbed to 100–200 cm^2 or even the whole side (see page 50). Such techniques can easily be devised for other products, where destructive sampling cannot be used.

The agar sausage technique (ten Cate, 1965) is another valuable sampling method, especially for checking the efficacy of the cleaning of work surfaces. It can be used only to assess relatively low numbers of microorganisms and is therefore inapplicable to most raw, cured meats.

Destructive sampling

Surface tissues

If it can be assumed that the microorganisms of interest are on the surface, as is the case, for example, with the development of slime on sides of bacon, the excision with aseptic precautions of areas of the external surfaces is a convenient method of sampling. With vacuum packed sliced bacon, of course, the slices can be taken, though there are obvious advantages in standardizing the proportions of fat and lean in the sample by excising equal areas of both tissues to make a composite sample. The same technique can be used for vacuum packed sliced cooked meats. The practice of selecting only lean meat for examination, especially for the purpose of determining the composition of the microflora is strongly deprecated where fatty tissues make up a significant proportion of the product (e.g. bacon) and where there is no direct evidence that only the lean is abnormal.

Internal tissues

With some defects, such as bone taint, it is necessary to obtain a sample from within the tissues and to ensure that there is no contamination from the surface. The technique of G. C. Ingram (see Sharp, 1963) or the modification by Gardner and Carson (1967) based on painting the external surface with an antiseptic dye was devised for obtaining samples of muscle. Again, according to circumstances, a decision will have to be made as to what tissues should be examined. In the case of bone taint, samples from the musculature close to the bone and samples of marrow from within the bone should be taken.

Comminuted cured meats

With comminuted products such as salami, baconburgers or frankfurters, a known weight can be taken. In some circumstances it may still be useful to distinguish between samples from the surface and the centre, e.g. in instances of microbial slime formation or discoloration at the surface of cured sausages.

Canned cured meats

There are standard reference books (e.g. Hersom and Hulland, 1963) on the microbiological examination of canned cured meats. There is a need to examine the jelly or brine, the surface of the product, especially adjacent to the can seams, and the centre of the product, in order to identify the nature of any defect.

Specific microbiological defects

There are a number of specific microbiological defects (e.g. "greening" in cooked ham and certain sausages, "pink spots" on salted casings), where comparative microbiological examination is useful. Samples are taken of the defective, as well as of the apparently normal material, and the results assessed on a relative as well as an absolute basis.

In summary, the sampling of cured meats for microbiological analysis depends on the nature of the meat, the type of sample which will most accurately reflect the condition(s) under investigation, and the feasibility of using a destructive sampling method.

Preparation of Sample

Diluent

In an effort to match the salt contents of diluents and culture media to that of the cured meat so as to minimize possible osmotic damage, 4% (w/v) NaCl is usually added to both (see Ghijsen, Gibbons, Hornsey and Riemann, 1958). It has been established that the protective effect of 0·1% (w/v) peptone in diluents (Straka and Stokes, 1957) applies also in the presence of salt (Patterson and Cassells, 1963). Thus the following diluent is recommended (% w/v): peptone, 0·1 and NaCl, 4·0.

Maceration

Samples of internal tissues and weighed samples of comminuted cured meats are conveniently prepared for dilution and plating by macerating in diluent (volume = 9 × the weight of the meat) in a mechanical blender. Two bottom-drive units (Warring blender type) and 2 top-drive units were compared with grinding with a pestle in a mortar with sand (Barraud, Kitchell, Labots, Reuter and Simonsen, 1967) as part of the programme of the International Organization for Standardization (ISO). Differences between counts of fresh and cured meats macerated by each method, though statistically significant, were small, but grinding with sand gave unacceptably low counts with some samples; also, standardization of a manual operation such as grinding is difficult and the method is not recommended. Results obtained with ham and salami are given in Table 1.

TABLE 1. A comparison of the counts obtained on samples of ham and salami macerated in different ways*

Method of maceration		Counts (%)† on	
Apparatus	Operation (sec/rpm)	Salami	Ham
MSE Atomix	30/6000 +60/12,000	100	100
Biorex	30/45,000	100	85
Ultra-Turrax	30/23,000	81	91
MSE Nelco	120/8000	67	89
Pestle and mortar (with sand)	120 sec	75	83

* Data of Barraud et al. (1967).

† Mean counts by all other methods expressed as percentages of the Atomix count.

Blenders commonly used operate over the range 6000–45,000 rpm and are standardized in use on the basis of 15,000–20,000 cutting revolutions. The duration of blending never exceeds 2·5 min and may be as short as 20 sec, thus minimizing the risk of overheating. Blender jars can be washed and autoclaved for re-use or sterilized by the method given below for the Ultra-Turrax, though if they are of the bottom-drive type, the cutter bearings need regular lubrication. It is recommended that a special box spanner is made to facilitate removal of the cutter unit for cleaning and lubrication.

The homogenizers made by Silverson (Silverson Machines, London) and Ultra-Turrax (Orme Scientific, Middleton, Manchester) work on a different principle and optimum conditions for use have yet to be determined. They may also be difficult to sterilize, but the following procedure has been found to be effective in preparing the Ultra-Turrax for immediate re-use:

(a) wipe clean immediately after use,

(b) operate for 5 sec at full power in warm water containing a detergent (e.g. Teepol),

(c) wipe with a clean tissue,

(d) repeat the operation (b) in industrial alcohol and allow the cutting blades to remain soaking in alcohol for c. 1 min before removing, and

(e) burn off residual alcohol and allow to cool before use.

Even after preparation of meat samples with counts higher than 10^8/ml (macerate), cleaning in the way prescribed reduced survivors to below the limits of detection of the counting procedure used.

Mechanical shaking

This method of removing organisms from the sample avoids the difficulty of sterilizing mechanical devices and is suited to the processing of large numbers of samples. It has been reported (Kitchell in Jayne-Williams, 1963) that shaking with glass beads consistently gives higher counts than macerating mechanically. Though the differences are small and of no account quantitatively, they may be important in flora studies, because the reduction appears to occur in the numbers of specific bacteria, viz. Gram negative rods. The Griffin flask shaker (Griffin and George Ltd., Wembley, England) has been shown to be a convenient means of mechanical shaking. However, to standardize the procedure, the machine should be calibrated for known loads (Fig. 1). In addition, information should be obtained on the duration of shaking required to give maximum counts. Under the conditions specified (Fig. 1) shaking for 3 min at 1000 oscillations/min is satisfactory for sliced bacon. The aim should be strict standardization, to reduce errors resulting from technique.

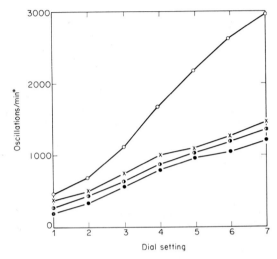

FIG. 1. Calibration curves of Griffin Flask Shaker for different numbers of 100 ml plastic bottles containing 20 ml diluent. No. of bottles in load: no load, (O); 2 bottles, (×); 4 bottles, (◗), and 8 bottles, (●).   * Rate of oscillation determined with a stroboscope.

Viable Counting and Identification

To avoid any possible lethal effect of hot agar on cells and to facilitate the isolation of cultures for identification, the pour plate technique should be avoided and all viable counts done by surface plating (see Barraud *et al.*, 1967).

There are a number of devices which can be used to transfer a known volume of each dilution to the surface of dried agar plates. For example:

(a) stainless steel cannula dropping pipettes (Astell Cat. No. 852),

(b) Pasteur pipettes (Harshaw Chemicals Ltd., Daventry, England),

(c) calibrated wire loops, and

(d) Davis 0·02 ml platinum tipped dropping pipettes (Astell Cat. No. 851).

For routine counting, 3 drops of each of 3 dilutions or spread drops of 4 dilutions can easily be accommodated on one Petri dish (diam, 9 cm). Where well isolated colonies are required, as in flora analyses, it is best to deliver a larger volume, usually 0·1 ml, of each dilution to a plate and spread the inoculum over the entire surface with a platinum wire or bent glass rod.

A summary of the most commonly used media in the microbiological analysis of cured meats is given in Table 2. Choice of media will depend on the nature of the cured meat and the purpose for which the sample was taken, i.e. an examination for spoilage types, potential foodborne pathogens, or indicator species.

TABLE 2. Media for microbiological examination of cured meats

Group	Medium	Incubation Temperature (°)	Time (days)
Total viable count	Plate count agar (PCA)	22	5
	(Oxoid) with and without salt*	30 or	3
Vibrio	Medium of Gardner (impress)	22	3
Aciduric lactic acid bacteria	Acetate agar (Rogosa, Mitchell and Wiseman, 1951) under H_2/CO_2 (95%/5%)	30	3
Enterococci	Medium of Barnes (1956) or	37	1
	Packer (1943)	37	2
Yeasts and moulds	Malt Extract agar, pH 3·5 (Oxoid)	25	3–5
Microbacterium thermosphactum	Medium of Gardner (1966)	22	2
H_2O_2 producing lactic acid bacteria	Manganese dioxide agar (Shipp, 1964)	30	3
Salmonella *Staphylococcus* *Clostridium*	See Thatcher and Clark (1968)		

* The level of NaCl should be close to that in the product; useful concentrations are 2, 4 and 10% (w/v).

The Microflora of Cured Meats

It is a common observation with vacuum packed meat, in contrast to the behaviour of meat exposed to air, that the maximum viable count is reached some time before off odours and flavours develop in the product. Thus counts alone may give no indication of the true state of the product, and for any useful conclusion to be drawn from the examination of the material it would be necessary to establish the composition of the microflora. A simple scheme for the purpose is given in Table 3 together with a list of commonly found spoilage microorganisms of some cured meats (Table 4).

TABLE. 3. Simplified scheme for identification of microorganisms commonly found on cured meats

Gram reaction*	Morphology	Catalase†	Oxidase‡	Presumptive groups	Schemes and methods for classification
+	Rods, cocci	−	NA	Lactic acid bacteria	Cavett (1963), Sharpe, Fryer and Smith (1966)
+	Non-sporing rods	+	NA	*Microbacterium thermosphactum*	McLean and Sulzbacher (1953), Davidson, Mobbs and Stubbs (1968)
−	Rods or cocco-bacilli	+	+	Pseudomonadaceae	Harrigan and McCance (1966)
			−	Enterobacteriaceae	Harrigan and McCance (1966)
−	Pleomorphic rods	+	+	*Vibrio*	Gardner (1971)
+	Cocci	+	NA	Micrococcaceae	Baird-Parker (1966)
+	Ovoid or spherical cells showing "budding"	+	NA	Yeasts	Beech, Davenport, Goswell and Burnett (1968)

* The Gram reaction and morphology are studied by making stained preparations from overnight cultures on PCA with 4% NaCl at 22°.
† The presence of catalase is determined by the production of gas, when a loopful of cells is emulsified in a drop of 10 vol H_2O_2 on a slide.
‡ Test of Kovacs (1956). NA, test not appropriate.

TABLE 4. Common spoilage microorganisms of cured meats

Type of cured meat	Organisms frequently found in spoilage flora
Bacon sides, unwrapped joints	Micrococcaceae, *Vibrio*, *Acinetobacter*, yeasts
Vacuum packed bacon, baconburgers, salami	Lactobacillaceae, Micrococcaceae, yeasts
Vacuum packed cooked cured meats	Lactobacillaceae, *Vibrio*, *Microbacterium thermosphactum*, Enterobacteriaceae
Canned hams	Lactobacillaceae
Discoloured (green) cooked cured meats	H_2O_2 producing Lactobacillaceae
Internal taints	*Vibrio*, Lactobacillaceae, Micrococcaceae, Enterobacteriaceae, *Clostridium*

References

BAIRD-PARKER, A. C. (1966). Methods for classifying staphylococci and micrococci. In *Identification Methods for Microbiologists*. Part A (B. M. Gibbs and F. A. Skinner, eds). London and New York: Academic Press.

BARNES, E. M. (1956). Methods for the isolation of faecal streptococci (Lancefield Group D) from bacon factories. *J. appl. Bact.*, **19**, 193.

BARRAUD, C., KITCHELL, A. G., LABOTS, G., REUTER, G. & SIMONSEN, B. (1967). Standardisierung der aeroben Gesamtkeimzahlbestimmung in Fleisch und Fleischerzeugnissen. *Fleischwirtschaft*, **47**, 1313.

BEECH, F. W., DAVENPORT, R. R., GOSWELL, R. W. & BURNETT, J. K. (1968). Two simplified schemes for identifying yeast cultures. In *Identification Methods for Microbiologists*. Part B (B. M. Gibbs and D. A. Shapton, eds). London and New York: Academic Press.

TEN CATE, L. (1965). A note on a simple and rapid method of bacteriological sampling by means of agar sausages. *J. appl. Bact.*, **28**, 221.

CAVETT, J. J. (1963). A diagnostic key for identifying the lactic acid bacteria of vacuum packed bacon. *J. appl. Bact.*, **26**, 453.

DAVIDSON, C. M., MOBBS, P. & STUBBS, J. M. (1968). Some morphological and physiological properties of *Microbacterium thermosphactum. J. appl. Bact*, **31**, 551.

ELLIOTT, R. J. & PORTER, A. G. (1971). A rapid cadmium reduction method for the determination of nitrate in bacon and curing brines. *Analyst, Lond.*, **96**, 522.

FOLLETT, M. J. & RATCLIFF, P. W. (1963). Determination of nitrite and nitrate in meat products. *J. Sci. Fd Agric.*, **14**, 138.

GARDNER, G. A. (1966). A selective medium for the enumeration of *Microbacterium thermosphactum* in meat and meat products. *J. appl. Bact.*, **29**, 455.

GARDNER, G. A. (1971). Microbiological and chemical changes in lean Wiltshire bacon during aerobic storage. *J. appl. Bact.*, **34**, 645.

GARDNER, G. A. (in press). A selective medium for the enumeration of salt requiring *Vibrio* spp from Wiltshire bacon and curing brines. *J. appl. Bact.*, **36**.

GARDNER, G. A. & CARSON, A. W. (1967). Relationship between carbon dioxide production and growth of pure strains of bacteria on porcine muscle. *J. appl. Bact.*, **30**, 500.

GARDNER, G. A. & PATTON, J. (1969). Variations in the composition of the flora on a Wiltshire cured bacon side. *J. Fd Technol.*, **4**, 125.

GHIJSEN, J. J., GIBBONS, N. E., HORNSEY, H. C. & RIEMANN, H. (1958). General bacteriology of beef and bacon curing brines. In *2nd International Symposium of Food Microbiology, Cambridge*. London: HMSO.

HARRIGAN, W. F. & MCCANCE, M. E. (1966). *Laboratory Methods in Microbiology*. London and New York: Academic Press.

HERSOM, A. C. & HULLAND, E. D. (1963). *Canned Foods. An Introduction to their Microbiology (Baumgartner)*. 5th Ed. London: J. A. Churchill.

INGRAM, M. (1960). Bacterial multiplication in packed Wiltshire bacon. *J. appl. Bact.*, **23**, 205.

JAYNE-WILLIAMS, D. J. (1963). Report of a discussion on the effect of the diluent on the recovery of bacteria. *J. appl. Bact.*, **26**, 398.

KOVACS, N. (1956). Identification of *Pseudomonas pyocyanea* by the oxidase reaction. *Nature, Lond.*, **178**, 703.

McLean, R. A. & Sulzbacher, W. L. (1953). *Microbacterium thermosphactum,* spec. nov., a non-heat resistant bacterium from fresh pork sausage. *J. Bact.,* **65,** 428.

Packer, A. R. (1943). The use of sodium azide (NaN₃) and crystal violet in a selective medium for streptococci and *Erysipelothrix rhusiopathiae. J. Bact.,* **46,** 343.

Patterson, J. T. & Cassells, J. A. (1963). An examination of the value of adding peptone to diluents used on the bacteriological testing of bacon curing brines. *J. appl. Bact.,* **26,** 493.

Rogosa, M., Mitchell, J. A. & Wiseman, R. F. (1951). A selective medium for the isolation and enumeration of oral and fecal lactobacilli. *J. Bact.,* **62,** 132.

Schonherz, Z. (1955). The determination of salt in brines, pickles, salted meats and seasonings by titration with mercuric nitrate solution. *Fd Mf.,* **30,** 460.

Sharp, J. G. (1963). Aseptic autolysis in rabbit and bovine muscle during storage at 37°. *J. Sci. Fd Agric.,* **14,** 468.

Sharpe, M. E., Fryer, R. F. & Smith, D. G. (1966). Identification of the lactic acid bacteria. In *Identification Methods for Microbiologists.* Part A (B. M. Gibbs and F. A. Skinner, eds). London and New York: Academic Press.

Shipp, H. L. (1964). *The green discoloration of cooked cured meats of bacterial origin.* Technical Circular No. 266. Leatherhead, Surrey: BFMIRA.

Straka, R. P. & Stokes, J. L. (1957). Rapid destruction of bacteria in commonly used diluents and its elimination. *Appl. Microbiol.,* **5,** 21.

Thatcher, F. S. & Clark, D. S. (1968). *Microorganisms in foods: their significance and Methods of Enumeration.* Toronto: University of Toronto Press.

Tofte Jespersen, N. J. & Riemann, H. (1958). The numbers of salt-tolerant bacteria in curing brine and on bacon. In *2nd International Symposium of Food Microbiology, Cambridge.* London: HMSO.

Routine Microbiological Examination of Wiltshire Bacon Curing Brines

G. A. GARDNER

Ulster Curers' Association, 2 Greenwood Avenue, Belfast BT4 3JL, Northern Ireland

Butchered sides of pork are pumped with a freshly prepared salts solution, *injection brine*. Each side receives 25 injections ("stitches") of brine, mostly in the shoulder and gammon. The pumped sides are then immersed in a *cover brine* for 3–5 days, which cures the unpumped surface and shallow tissues such as belly. The cover brine is repeatedly used, being filtered and reconstituted with curing salts between each curing cycle, Fig. 1. The

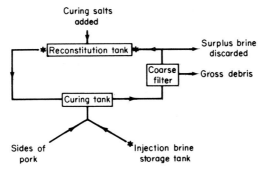

FIG. 1. Wiltshire curing brine cycle. *Sampling points for routine microbiological and chemical quality control of curing brines.

temperature range for curing is 3·3–5·6° and the range in concentration of curing salts in both injection and cover brines is as follows (% w/v): NaCl, 26–29; $NaNO_3$, 0·1–0·25; and $NaNO_2$, 0·05–0·07. Normally a cover brine will have a pH of 6·2–6·5.

In the curing industry the cover brine is regarded as an important material in itself, with good management it can be used indefinitely. Brine stability is directly related to microbiological growth and activity; this activity is measured in terms of the reduction of nitrate and/or nitrite with the associated increase in pH. It is for this reason that the routine microbiological and chemical examinations of brines from all bacon factories in Northern Ireland were initiated in 1960 and have since become an important part of the quality control organization of the industry. The laboratory examination of curing brines and its value in detecting defects in the production cycle are outlined in this chapter.

Sampling

Injection brine should be sampled from the preparation or storage tank (Fig. 2). Cover brine should be sampled only in the reconstitution tank

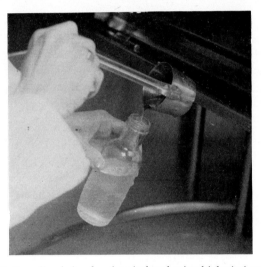

FIG. 2. Sampling brine for chemical and microbiological analysis.

with the mixing device operating so that the brine is well mixed. Samples should be taken before (*spent pickle*) and after (*reconstituted*) the addition of curing salts (Fig. 1). Samples are transported to the laboratory in an insulated box containing an ice pack. Any sample arriving at the laboratory with a temperature greater than 7° is not regarded as suitable for testing, as changes in the brine could take place during transit, and thus the sample would not reflect accurately its condition at the factory.

Routine Analyses

Chemical

The routine analysis of brines for pH, sodium chloride (Schonherz, 1955), sodium nitrate and sodium nitrite (Elliott and Porter, 1971), is useful for two purposes. Firstly, it is a check that the correct quantities of these salts are used in the preparation and reconstitution of brines, and secondly it provides data on the stability of cover brines. Gross microbiological activity in cover brine can result in the reduction of nitrate to nitrite and also the reduction of nitrite. Routine data can indicate the type of treatment necessary to control the brine and subsequently the cure of the bacon.

Microbiological

The tests employed in the routine examination of brines are given in Table 1.

TABLE 1. Microbiological tests for routine examination of bacon curing brines

Test	Injection brine	Cover brine
Direct microscopic count	−*	+
Count on nutrient agar (4% (w/v) NaCl) at 22°	+	+
Escherichia coli I	−	+
Vibrio	−	+

* Test: done (+), not done (−).

Direct Microscopic Count (DMC)

The enumeration of the total halophilic bacterial population in brines by viable counts is time consuming, and the DMC test (Ingram, 1958) was adopted so that gross changes in cover brines would be quickly recognized, thus minimizing the delay in treating an unstable brine—for additional details, see page 29.

Method. Shake the sample thoroughly. Transfer 1 loopful to a clean dry counting chamber such as Helber haemocytometer; cover with a glass cover slip and examine by dark ground or phase contrast illumination at a magnification of *c.* × 500. Slides after use are rinsed in cold water and stored in industrial alcohol. Immediately before use they are dried with a clean dry cloth.

The first indication of a brine becoming unstable is an increase in the DMC (Eddy and Kitchell, 1961). This is followed by a rise in the nitrite, resulting from nitrate reduction, and this usually becomes evident when the DMC exceeds 100×10^6 cells/ml. Data from two brines, one stable and the other unstable, illustrating this point, are given in Fig. 3. These data are derived from routine factory samples submitted for routine examination over a period of 11 months.

One of the main causes of sudden bacterial growth in brine is a temperature rise to 7° and above. The normal curing temperature range is 3·3–5·6° and within this range there is no noticeable increase in the bacterial load. The rise in temperature of the brine can be due to: (a) breakdown of cellar refrigeration, or (b) the temperature of pork sides going into cure being too high. Also, if the NaCl level in the brine falls to less than 20% (w v), there is more rapid multiplication of the brine microflora.

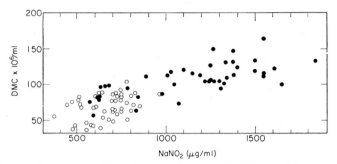

FIG. 3. Relationship between direct microscopic count (DMC) and sodium nitrite in two cover brines. Open circles, stable brine, and closed circles, unstable brine.

As there is now a limit of 200 ppm nitrite in bacon (Anon, 1971), it is imperative to control the levels in the cover brines, as these are responsible for curing the surface and shallow tissues of the bacon side. In addition, high levels of nitrite can result in brown or green discolorations ("nitrite burn") of the meat. Conversely, the absence of nitrate and nitrite in the cover brine may cause "miscuring" of these tissues, i.e. no source of NO for converting the meat pigment to the nitroso form.

Count on Nutrient Agar with 4% (w/v) NaCl (NA4)

Viable counts on nutrient agar containing 4% NaCl (NA4) are used to indicate the level of bacteria which may ultimately grow on the bacon. These organisms do not grow, but can survive for a long time in the brine. The NA4 counts are useful in assessing the contamination contributed to the brine from the pork and equipment.

Method. Prepare serial dilutions of the brine in 4% (w/v) NaCl, 0·1% (w/v) peptone diluent, and plate out 0·1 ml by the drop and spread technique using Davis dropping pipettes (Astell Cat. 851). The composition of the medium used is as follows (% w/v): peptone (Oxoid L37), 1·0; Lab Lemco (Oxoid L29), 0·24; NaCl, 4·0; agar, 1·25; pH 7·6. Sterilized by autoclaving at 15 lb for 15 min. Colonies are enumerated after incubation of plates for 5 days/22°. Plates, dried at 37° overnight, should be stored in a refrigerator and used within 3–4 days of preparation.

These counts with freshly prepared injection pickles are generally $< 10^3$/ml, but with cover brines one can obtain counts as high as 10^6/ml or more. Highly contaminated injection pickle may reflect contamination of water and/or salts used in its manufacture or contamination acquired from tanks, pipes or even cover brine.

In cover brines, NA4 counts are indicative primarily of the microbiological condition of the pork sides coming into cure, which in turn reflects the general standard of hygiene and chilling in the factory.

It may be of interest to note that the curing of defrosted, frozen pork or the re-tanking of matured bacon sides results in a sharp increase in the NA4 counts in the brine. This raw material will have a much higher level of surface contamination than normal pork carcases and suggests that there is a transfer of bacteria from the meat to the brine. However, it is known that the brine can contribute to the NA4 counts on bacon sides (Gibbons and White, 1941; Ingram, 1958; Tofte Jespersen and Riemann, 1958), and it is bacteria in this group which will subsequently grow and cause spoilage of the product. These observations suggest that there is a balance between removal and gain of bacteria on the bacon side during immersion in the tank and that the direction of change most probably depends on the extent of the difference between the loads on the meat and the brine.

Count of Escherichia coli *I*

E. Coli type I are known to occur, but, though unable to grow, can survive for some time in curing brines (Garrard and Lochhead, 1939; Buttiaux and Moriamez, 1958; Patton, 1958). The number of these organisms in a cover brine can be used as an index of faecal pollution.

Method. One ml of brine is added to 9 ml sterile diluent (4% (w/v) NaCl, 0·1% (w/v) peptone) and filtered through a membrane (Oxoid 47 mm diam). The membranes are transferred to pads (Whatman Grade 17, 50 mm) in aluminium tins and soaked with Membrane Enriched Teepol broth (Oxoid MM369). These are incubated for 4 h/37° followed by 18–24 h/ 44° in a water bath. All flat yellow, shiny colonies should be counted. Pink or colourless colonies should be ignored.

The sudden appearance of *E. coli* in a brine can usually be related to poor practices on the slaughterline, e.g. puncturing of intestinal tract and subsequent contamination of the carcass and hence the brine. *E. coli* can survive for considerable periods in brine (Fig. 4).

Recent work has shown that this technique does not permit the enumeration of all viable *E. coli* (Fig. 4). The longer they are exposed to the brine, the poorer is their recovery. Work is at present being directed to the improvement of methods for recovery of these brine-damaged cells. However, regular analyses and survival data help in differentiating between recent contamination and the residual survivors from a more remote contamination.

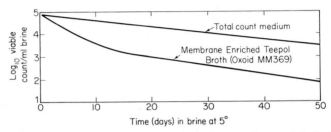

FIG. 4. Effect of recovery medium on the survival of pure cultures of *Escherichia coli* in Wiltshire bacon cover brine (Mean results of 14 strains).

Advisory Standards

For advisory purposes a set of standards has been based (Table 2) on the

TABLE 2. Microbiological tests and advisory standards for Wiltshire curing brines

Category	Injection brine Count on 4% (w/v) salt medium ($\times 10^3$/ml)	*E. coli I* (/ml)	Cover brine Count on 4% (w/v) salt medium ($\times 10^3$/ml)	Direct micro-scopic count ($\times 10^6$/ml)	*Vibrio** ($\times 10^3$/ml)
A Good	<0·5	<1	<50	<50	<1·0
B Fair	0·5–1·0	1–10	50–100	50–100	1·1–10·0
C Poor	1·1–5·0	11–100	101–500	101–150	10·1–100
D Very poor	<5·0	<100	<500	<150	<100

* See Gardner (in press).

tests described above. The lines of demarcation between the categories are in practice not used as rigidly as implied. In many cases one has to consider the situation in relation to specific conditions in a particular factory. It is important to consider changes in a brine rather than act on results from one sample. Also, microbiological findings must be assessed in conjunction with the chemical data, so that in the event of some defect or potential defect, remedial action can be swiftly taken.

All cover brines with a category D classification in the direct microscopic count need immediate attention. There are a number of alternatives, e.g. discarding the brine, dilution with freshly prepared injection brine, filtration, or acid treatment. However, brine stability is only an indicator of a fault elsewhere and unless this is isolated and identified, the problem will be of a recurrent nature.

References

ANON (1971). Food and Drugs; Composition and Labelling; Preservatives in Foods. S.R.&O.(NI). No. 161. Belfast: HMSO.

BUTTIAUX, R. & MORIAMEZ, J. (1958). Le comportement des germes tests de contamination fécale dans les saumures de viandes. In *2nd International Symposium of Food Microbiology, Cambridge*. London: HMSO.

EDDY, B. P. & KITCHELL, A. G. (1961). Nitrate and nitrite metabolism in a bacon curing brine and their relation to the bacterial population. *J. Sci. Food Agric.*, **12**, 146.

ELLIOTT, R. J. & PORTER, A. G. (1971). A rapid cadmium reduction method for the determination of nitrate in bacon and curing brines. *Analyst, Lond.*, **96**, 522.

GARDNER, G. A. (in press). A selective medium for the enumeration of salt requiring *Vibrio* spp from Wiltshire bacon and curing brine. *J. appl. Bact.*, **36**.

GARRARD, E. H. & LOCHHEAD, A. G. (1939). A study of bacteria contaminating sides for Wiltshire bacon with special consideration of their behaviour in concentrated salt solutions. *Can. J. Res. (D).*, **17**, 45.

GIBBONS, N. E. & WHITE, W. H. (1941). Canadian Wiltshire bacon. XV. Quantitative bacteriological and chemical changes in tank pickle and on bacon during cure and maturation. *Can. J. Res. (D).*, **19**, 61.

INGRAM, M. (1958). The general microbiology of bacon-curing brines, with special reference to methods of examination. In *2nd International Symposium of Food Microbiology, Cambridge*. London: HMSO.

PATTON, J. (1958). Observations of coliform bacteria in bacon-curing brines. In *2nd International Symposium of Food Microbiology, Cambridge*. London: HMSO.

SCHONHERZ, Z. (1955). The determination of salt in brines, pickles, salted meats and seasonings by titration with mercuric nitrate solution. *Fd Mf.*, **30**, 460.

TOFTE JESPERSEN, N. J. & RIEMANN. H. (1958). The numbers of salt-tolerant bacteria in curing brine and on bacon. In *2nd International Symposium of Food Microbiology, Cambridge*. London: HMSO.

Methods for the General Microbiological Examination of Wiltshire Bacon Curing Brines

A. H. VARNAM AND J. M. GRAINGER

Department of Microbiology, The University, Reading RG1 5AQ, England

The most recent widely available collection of publications on the general microbiology of bacon curing brines comprises a part of the Proceedings of the Second International Symposium on Food Microbiology which was held in 1957 (Eddy, 1958). For a number of years since then interest in curing brine microbiology has been limited to aspects which may be of use in routine quality control. The sampling procedures and routine analyses involved in such work are dealt with elsewhere in this publication (p. 21).

Recently however there has been some revival of interest in the general microbiological aspects. This contribution is concerned mainly with the procedures used in a study of a range of different groups of bacteria including the dominant ones from Wiltshire cure cover brine; in addition some of the results obtained in the study are outlined. Samples were taken over a three year period from three factories in the West Country of England. The dominant microorganisms, which have been little studied previously, were Gram negative rod-shaped oxidase-positive bacteria. The importance of this group of bacteria in controlling the major chemical changes that occur in the curing tank is indicated by the large numbers present and by their ability to grow and reduce nitrate under the prevailing environmental conditions. Members of the Staphylococcus-Micrococcus group represented usually about 10–20% of the bacteria recovered. They survive the curing process and some may grow slowly in the brine. Colonies of this group of bacteria develop on the media which are used in routine examination of brines and the counts obtained are taken to reflect factory hygiene standards. Small numbers of lactobacilli and leuconostocs were isolated; such bacteria from this source have been little studied previously. Other microorganisms may occur in bacon curing brines and in other types of meat curing brines. These include other bacteria, e.g. coliforms, salmonellae, endospore-forming bacteria, streptococci and pediococci (see Eddy, 1958; and p. 21), and yeasts (Mrak and Phaff, 1948).

The literature upon which we have been able to draw is limited mainly to published work in journals, symposium proceedings and books. We realize that there may be further information in those reports, technical bulletins, etc. which have limited circulation within the meat industry and of which we have no knowledge. We are grateful to those who have drawn our attention to some of this information.

We hope that the procedures to be described and discussed will be of some interest to those who work with cured meats and are concerned to learn more of the general microbiology of bacon curing brines. A further point of possible interest is that this environment is a source of "moderately" halophilic bacteria, microorganisms which have attracted much less attention than the in some ways more dramatic "extreme" halophiles.

Gram negative oxidase-positive rod-shaped bacteria and members of the Staphylococcus-Micrococcus group

These two groups of bacteria are numerically the most important microorganisms of Wiltshire bacon curing brines. Commonly 10^8–10^9 colonies /ml of cover brine may be recovered of which the great majority are Gram negative rod-shaped bacteria. Both groups of bacteria may be recovered by using one series of plating media, i.e. a common basal medium to which is added a range of concentrations of NaCl.

Enumeration

Plating medium

The basal plating medium is a nutrient agar supplemented with 15–25% (v/v) unheated aqueous extract of pork (see Appendix for method of preparation of pork extract). The nutrient agar which we use has the following composition (g/1): Lab Lemco (Oxoid), 5; peptone (Evans), 10; agar (Oxoid No. 3), 20; pH 6·8–7·0. This medium is similar to that used by Hornsey and Mallows (1954) for beef curing brines but nitrate is omitted as this constituent did not increase the colony count for samples which we examined. We have no experience with dehydrated nutrient agar but if it is necessary to use such a product it would be prudent to compare a number of different specifications before deciding upon one for routine use. Supplementation of media with pork extract for plating bacon curing brines increases the colony count. This procedure was used by Ingram, Kitchell and Ingram (1958) following the work of Hornsey and Mallows (1954) with beef curing brines. For the samples which we examined the inclusion of this constituent in plating media increased the count by a factor of at least 10 (Table 1). The effect was most marked with the medium containing

TABLE 1. Colony count of Wiltshire bacon curing brine B20: the effect of supplementing the plating medium* with 25 % (v/v) pork extract and a range of concentrations of NaCl

NaCl content of plating medium (% w/v)	10^{-6} × colony count/ml cover brine† on NaCl-containing plating medium		Incubation period (days)
	without pork extract	with pork extract	
0·5	0·4	nd	5
5	5	10	10
10	20	100	20
20	50	550	30

Dilutions prepared in liquid form of plating medium + NaCl (0·5, 5, 10 or 20%), plates inoculated by surface drop method and incubated at 25° in air.

* Plating medium (g/1): Lab Lemco, 5, peptone, 10; agar, 20; pH 6·8–7·0.

† Direct microscopic count of sample: 1100 × 10^6/ml. nd = not done.

20% NaCl on which the greatest colony counts were obtained, the colonies being mainly Gram negative rod-shaped bacteria (*vide infra*). At least one important function of pork extract may be to overcome inhibitory effects of the plating medium which affect the initiation of growth of many of the Gram negative rod-shaped bacteria especially in dilute inocula (A. H. Varnam and J. M. Grainger, unpubl. data); in its absence such bacteria either do not grow or grow poorly.

Among the bacteria of bacon curing brine there is represented a wide range of NaCl requirements for growth including "facultative" and "obligate" requirements and comprising an almost continuous gradation from non-halophiles to "moderate" halophiles *sensu* Larsen (1962) and Gibbons (1969). It should be noted that there is much confusion and disagreement on the definitions and terminology used for the different levels of requirement for NaCl; reference may be made to Ingram (1957), Larsen (1962), Kushner (1968) and Gibbons (1969) for discussion of the problem. Therefore in order to recover a widely representative range of isolates a series of plating media is used comprising the basal plating medium described previously with the addition of the following concentrations of NaCl (% w/v): 0·5, 5, 10 and 20. For the preparation of media containing 10 and 20% NaCl it is convenient to weigh into bottles appropriate amounts of NaCl, sterilize by hot air and add the NaCl to melted medium immediately before use.

We examined the identity of colonies which developed on each medium in the series of plating media. The proportion of colonies which were of Gram negative rod-shaped bacteria increased with the increase in NaCl content of the medium and *vice versa* for the Staphylococcus-Micrococcus group. Thus with the medium containing 0·5% NaCl the cocci were dominant whereas with those containing 10 and 20% NaCl, which recovered the greatest number of colonies, the rod-shaped forms were dominant. For the medium containing 5% NaCl there was no consistent pattern in the samples which we examined. Ingram (1958) suggested that the total colony count of bacon curing brines may approximate to the sum of the counts on the media containing 10 and 20% NaCl. In this way estimates of colony count which approach the direct microscopic count may be obtained (see Table 1) and Gram negative rod-shaped bacteria may be shown commonly to comprise 80–90% of the bacteria recovered. Reference should be made to Gardner (p. 21) for the use of colony counts on a 4% NaCl medium as indicators of standards of factory hygiene. In pure culture studies we recorded among the Gram negative rod-shaped bacteria a range of minimum requirements for growth from zero to 10% added NaCl, many isolates being capable of growth in concentrations that are present in cover brine. In contrast for the Staphylococcus-Micrococcus group no isolate had an "obligate" requirement for NaCl and many did not grow in media containing more than 15% NaCl.

Diluent

For each plating medium the diluent is its liquid equivalent but without pork extract. Dilutions of sample must be plated without delay. The use of diluent containing the same NaCl concentration as that of the plating medium follows recommended practice for bacon curing brines (Eddy, 1958, p. 333), the purpose being to reduce losses in viability which may occur when cells are suspended in inimical concentrations of NaCl. In this context the comment of Gibbons (1969) that "one must think 'salt' at all times" when working with halophiles is an important maxim to remember. The immediate plating of dilutions and the incorporation of a nutrient medium in the diluent are further attempts to protect viability. The value of adding a low concentration of peptone to diluents is widely appreciated —see Jayne-Williams (1963) for a report on a discussion on diluents— and Patterson and Cassells (1963) demonstrated the value of this procedure for bacon curing brines plated on a 4% NaCl medium.

Plating method

Samples are plated by a surface drop method similar to that of Miles and Misra (1938). In our experience surface plating methods gave greater colony counts than the pour plate method with media containing 5 and 10% NaCl. The increase was 2–4 fold and was more marked with the 10 than with the 5% NaCl-containing medium. We have not made this comparison for 20% NaCl because it is unwise to use the pour plate method with a medium having such a high temperature of solidification. Of the two surface plating methods we chose the drop method which gave 3–4 fold greater colony counts than the spread plate method with media containing 10 and 20% NaCl, the difference being most marked with the higher NaCl concentration. The reason for the increased count with the drop method may be that inhibitory effects of media to which many of the Gram negative rod-shaped bacteria isolated appear to be sensitive are overcome more successfully when cells are close together than when well separated.

Incubation conditions

Incubation is in air at 25°; at 17° colony development is slower than at 25° but the final colony counts at the two temperatures are usually similar. A different incubation period is used for each concentration of NaCl in the plating medium, i.e. for the medium containing 0·5% NaCl the incubation period is 5 days after which no additional colonies appear, whereas for the media containing 5, 10 and 20% NaCl the incubation period is, respectively, 10, 20 and 30 days (Table 2). Plates should be examined with

TABLE 2. Colony count of Wiltshire bacon curing brine B20: the effect of incubation period using a plating medium* containing a range of concentrations of NaCl

NaCl content of plating medium (% w/v)	$10^{-6} \times$ colony count/ml cover brine on NaCl-containing plating medium Incubation period (days)						
	5	10	15	20	25	30	35
0·5	0·4	—	—	—	—	—	—
5	1·5	5	—	—	—	—	—
10	0·8	4	75	100	—	—	—
20	nc	2·5	15	300	450	550	—

* Plating medium: as in Table 1 + pork extract (25% v/v); — = no further increase in number of colonies, and nc = no colonies developed.

the aid of a hand lens or low-power magnification of a microscope so that the small slowly-growing colonies are not overlooked. Precautions to avoid evaporation of medium during prolonged incubation must be taken, e.g. by enclosing the plates in a closed polythene bag containing moistened absorbent cotton wool.

Isolation and purification

Freshly isolated cultures should be maintained initially on a medium of the same composition as that on which they were isolated. When details of requirements for NaCl and pork extract are known the cultivation medium may be modified accordingly. It is not necessary to add pork extract to broth media.

Purification of cultures of the Staphylococcus-Micrococcus group may be achieved by the usual method of repeatedly picking a single colony into broth, incubating and re-streaking. For the Gram negative rod-shaped bacteria, however, this procedure may not be adequate and in some instance considerable difficulty may be experienced in obtaining pure cultures. The problem lies in the fact that a medium containing a high concentration of NaCl is a selective medium. Thus it is possible to carry over in the inoculum unwanted organisms which either are unable to grow or grow very slowly. This situation is well illustrated when a medium containing 20% NaCl is used to purify a culture of a Gram negative rod-shaped bacterium in which there are cells of a micrococci that cannot grow at that concentration of NaCl (Fig. 1). The incipient contamination may remain undetected by microscopic examination but may be revealed when the culture is streaked on a non-selective medium, i.e. one containing a suitably low concentration of NaCl (Fig. 1). During storage and subculture hitherto undetected micrococci cells may increase in number sufficiently to become detectable microscopically. Thus cultures should be streaked frequently on suitable media and examined microscopically to test for the presence of contaminants. When there is much difficulty in eliminating contaminating micrococci other procedures may be necessary, e.g. the use of a medium containing penicillin and a suitably low concentration of NaCl. The problem of ensuring purity of cultures is one which is not commonly discussed in published work. Ingram (1957) commented on such difficulties for halophilic bacteria and, in a more general context, Cowan's (1970) fourth and seventh Principles of Heretical Taxonomy are a necessary reminder that too much attention cannot be given to the problem.

Identification

For examination of cell form and arrangement it is not enough to rely solely on the use of heat-fixed smears prepared from cultures on media containing high concentrations of NaCl and stained by Gram's method. Although the problems caused by the precipitation of NaCl in the smear may be reduced by fixing with, e.g. acetic acid (Dussault, 1955), and by

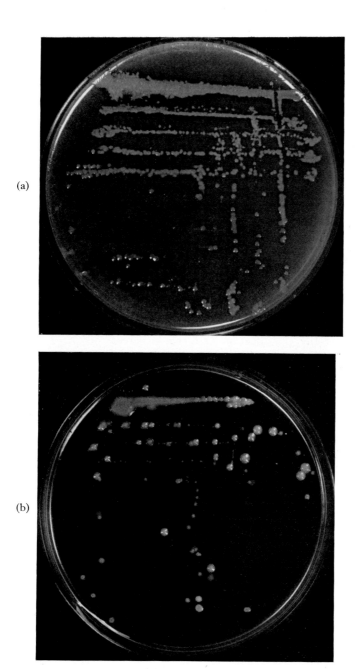

(a)

(b)

F IG. 1. A stage in the purification of a culture of a Gram negative rod-shaped bacterium from Wiltshire bacon curing brine showing the appearance of streak plates of media containing: (a) 20% NaCl on which there are colonies of the Gram negative bacterium, and (b) 5% NaCl on which there are in addition colonies of micrococci, i.e. the large colonies.

using cultures grown on the surface of agar media the results are not always wholly satisfactory. The examination of wet preparations by phase-contrast microscopy is a most valuable supplement to staining procedures.

For the identification of members of the Staphylococcus-Micrococcus group the scheme of Shaw, Stitt and Cowan (1951) in which one genus, *Staphylococcus*, is recognized has been used for isolates from bacon curing brines, e.g. Kitchell (1958); Patterson (1966). In contrast Baird-Parker (1965) used a scheme (Baird-Parker, 1962, 1963) in which the genera *Staphylococcus* and *Micrococcus* are recognized. Using the Baird-Parker scheme and a method and medium described by Baird-Parker (1966) based on that proposed by the International Sub-committee on Staphylococci and Micrococci (Anon, 1965) we identified 41 out of 49 such isolates as members of the genus *Micrococcus*.

Published work on the non-halophilic and "moderately" halophilic Gram negative rod-shaped bacteria of bacon curing brine is insufficient for their taxonomic position to be clearly established. Indeed very little is known about "moderately" halophilic bacteria in general. The published schemes for the classification of Gram negative rod-shaped bacteria are based on studies which have not included an examination of cultures with special NaCl requirements for growth. We have not attempted to use these schemes for bacteria from bacon curing brines because the relationships between "moderately" halophilic and non-halophilic forms are so poorly understood. It is not appropriate here to discuss in any detail the taxonomy of the representative cultures which we have studied but a brief mention of some features may be of interest. A range of different morphological forms was encountered including straight-sided, curved and irregular rods but as this feature may change with cultural conditions, particularly the concentration of NaCl present, and to change after first isolation we placed little weight on it for characterization of isolates. All 108 isolates were motile. The flagellar pattern appeared to be polar using Rhodes' (1958) modification of Fontana's silver impregnation method but it was particularly difficult to interpret the results in preparations from a medium containing a high concentration of NaCl. Some biochemical features were influenced by the concentration of NaCl in the test medium, e.g. nitrate reduction, hydrolysis of gelatin, starch and fat (see also Garrard and Lochhead, 1939; Ingram, 1957; Gibbons, 1958). Therefore it is important to make observations at more than one concentration of NaCl and to describe adequately the conditions under which tests are done. All cultures were oxidase-positive by the method of Kovacs (1956). For the majority of cultures (73 out of 108) there was no detectable acid production from glucose when tested by the method of Hugh and Leifson (1953). Twenty-four isolates produced acid from glucose oxidatively and 11 isolates produced acid but no gas fermentatively. Two named cultures of "*Vibrio costicolus*" also produced acid but no gas fermentatively and corresponded in other features to the small number of our isolates which gave this reaction. There was no clear relationship between biochemical features and NaCl requirements for optimum growth of isolates but in general the biochemically relatively inactive cultures, comprising the majority, were isolated on the 20% NaCl plating medium and had a maximum NaCl concentration for growth in excess of 25%.

Lactobacillus, Leuconostoc and Pediococcus

Lactobacilli and leuconostocs form a small part of the microflora of Wiltshire bacon curing brines. In our experience colony counts of 10^2–10^3/ml of cover brine are usual although we have recorded occasional counts of 10^5 and 10^6/ml and pediococci occur very rarely (Varnam and Grainger, 1972). There is little mention elsewhere in the literature of the isolation of members of these three genera from bacon curing brines. Ingram (1958) suspected that certain Gram positive rod-shaped bacteria were lactobacilli and Spencer (1967) reported only on the proportions of the different species of the genera.

Enumeration

Plating medium

A Tween-containing medium is used; for bacon curing brines acetate is not incorporated in the plating medium. The selective action of media originally developed for the enumeration of these bacteria from man and hamsters (Rogosa, Mitchell and Wiseman, 1951) and from grass and silage (Keddie, 1951) depends upon the presence of a 0·2 molar acetate buffer, pH 5·4. Acetate-containing media are used successfully for this purpose with a wide range of sources but such media are not suitable for bacon curing brines; commonly either no colonies or few colonies develop on Keddie's Acetate Agar medium—see Harrigan and McCance (1966) for composition —and a similar effect occurs with Rogosa's medium. However by omitting the acetate buffer a suitable plating medium is obtained (Varnam and Grainger, 1970). Two such suitable media are Keddie's Tween Agar medium, i.e. the "basal medium" of the Acetate Agar medium, which we use, and the APT (all purpose Tween) medium of Evans and Niven (1951). The basis of the method is that in bacon curing brines, unlike many other sources, there are very few if any other microorganisms capable of growth on such media of pH 5·4 when incubation is in a strictly anaerobic atmosphere (see *Incubation conditions*). It should be noted that dehydrated

c

APT medium (Difco and BBL) has an unadjusted pH value of 6·7 which should be adjusted to 5·4 when the medium is used for selective enumeration purposes. Plate Count Agar medium (we have used only the Oxoid product, adjusted to pH 5·4) is less satisfactory than Tween-containing media because colonies of unwanted bacteria may develop. The inclusion of 5% (w/v) or more NaCl in the plating medium reduces the colony count, an observation made also by Deibel and Niven (1958) with American ham curing brines.

Diluent

Following the principle of using a liquid equivalent of the plating medium the diluent is Keddie's Tween broth medium which is similar in composition to the solid medium but agar is omitted and the pH value is 6·2 (R. M. Keddie, pers. comm.).

Plating method

The surface drop method is used. In a comparison of different plating methods we noted that surface plating methods gave colony counts up to threefold greater than those obtained with the pour plate method. For enumeration purposes we chose the drop method for reasons of economy of time and materials.

Incubation conditions

Incubation is at 25° for 6 days in an anaerobic jar fitted with a room-temperature catalyst and filled with an atmosphere of hydrogen (95% v/v) and carbon dioxide (5% v/v). Similar colony counts are obtained at 30°. The gaseous environment for incubation must be strictly anaerobic otherwise unwanted colonies may develop. For instance in an atmosphere of nitrogen and carbon dioxide colonies of micrococci are able to grow and these may inhibit the lactic acid bacteria.

Isolation and purification

Keddie's Tween media are suitable for routine cultivation of isolates i.e. Tween agar (pH 6·2), Tween broth and Tween semi-solid agar (similar in composition to the broth medium but with the addition of agar, 1g/1, and K_2HPO_4, 5g/1; R. M. Keddie, pers. comm.). Also suitable are APT medium and the MRS medium of de Man, Rogosa and Sharpe (1960). During the purification of cultures incubation of streak plates in air and subculture into nutrient broth aid in the detection of contaminating micrococci.

Identification

A preliminary screening of isolates for presumptive allocation to the genera *Lactobacillus*, *Leuconostoc* and *Pediococcus* is achieved by examining the following three features in 24 h cultures in Tween semi-solid agar medium inoculated directly from colonies on isolation plates: (a) cell shape and arrangement and Gram reaction using heat-fixed stained smears and careful examination of wet preparations by phase-contrast microscopy; (b) homofermentative and heterofermentative activity by plunging into the culture a red-hot wire loop, the evolution of copious amounts of gas bubbles indicating heterofermentative activity (R. M. Keddie, pers. comm.), and (c) catalase activity by adding to a pellet of centrifuged cells a few drops of hydrogen peroxide. Following purification cultures are identified by the procedures described by Sharpe, Fryer and Smith (1966).

We examined more than 50 colonies from each of 8 samples of cover brine from different factories by the preliminary screening method desscribed previously. Lactobacilli and leuconostocs occurred in all samples; in five samples members of the two genera occurred in approximately equal proportions. All lactobacilli examined in this way were homofermentative. Identification of a small number of representative cultures showed that all of 29 lactobacilli were streptobacteria (i.e. homofermentative, grew at 15°) but their characteristics did not correspond exactly to previously described species; of 18 leuconostocs, 10 were strains of *Leuconostoc mesenteroides*. One isolate was a pediococcus and this was identified as a strain of *Pediococcus cerevisiae*.

Appendix

Pork extract is prepared by the following procedure which is based on that of Tofte Jespersen and Riemann (1958). Mix minced fresh lean fillet of pork (350g) with 200 ml demineralized/glass distilled water at 50° and maintain at that temperature for 1 h. Press the mixture in a cloth, collect the juice and make the volume up to 300 ml. Adjust to pH 6·5, strain several times through filter paper and sterilize by Seitz filtration. Add the extract, 15–25 % (v/v), immediately before use to medium previously melted and cooled to *c*. 50°.

Acknowledgements

We thank Miss Linda Janes for her conscientious technical assistance and Mr M. J. Crowder for taking the photographs. We gratefully acknowledge the willing co-operation of members of the cured meat industry.

References

ANON. (1965) Recommendations of International Subcommittee on taxonomy of Staphylococci and Micrococci *Int. Bull. bact. Nomencl. Taxon.*, **15,** 109.

BAIRD-PARKER, A. C. (1962). The occurrence and enumeration, according to a new classification, of micrococci and staphylococci in bacon and on human and pig skin. *J. appl. Bact.*, **25,** 352.

BAIRD-PARKER, A. C. (1963). A classification of micrococci and staphylococci based on physiological and biochemical tests. *J. gen. Microbiol.*, **30,** 409.

BAIRD-PARKER ,A. C. (1965). The classification of staphylococci and micrococci from world-wide sources. *J. gen. Microbiol.*, **38,** 363.

BAIRD-PARKER, A. C. (1966). Methods for classifying staphylococci and micrococci. In *Identification Methods for Microbiologists.* Part A, p. 59 (B. M. Gibbs and F. A. Skinner, eds). London and New York: Academic Press.

COWAN, S. T. (1970). Heretical taxonomy for bacteriologists. *J. gen. Microbiol.*, **61,** 145.

DEIBEL, R. H. & NIVEN, C. F. (1958). Studies on the microflora of commercial ham curing brines and their significance in curing. In *The Microbiology of Fish and Meat Curing Brines*, p. 149 (B. P. Eddy, ed). London: HMSO.

DUSSAULT, H. P. (1955). An improved technique for staining red halophilic bacteria. *J. Bact*, **70,** 484.

EDDY, B. P. (1958) (ed.). *The Microbiology of Fish and Meat Curing Brines*. Proceedings of the Second International Symposium on Food Microbiology. London: HMSO.

EVANS, J. B. & NIVEN, C. F. (1951). Nutrition of the heterofermentative lactobacilli that cause greening of cured meat products. *J. Bact.*, **62,** 599.

GARRARD, E. H. & LOCHHEAD, A. G. (1939). A study of bacteria contaminating sides for Wiltshire bacon with special consideration of their behaviour in concentrated salt conditions. *Can. J. Res.*, D **17,** 45.

GIBBONS, N. E. (1958). The effect of salt on the metabolism of halophilic bacteria. In *The Microbiology of Fish and Meat Curing Brines*, p. 69 (B. P. Eddy, ed.). London: HMSO.

GIBBONS, N. E. (1969). Isolation, growth and requirements of halophilic bacteria. In *Methods in Microbiology*, Vol. 3B, p. 169 (J. R. Norris and D. W. Ribbons, eds). London and New York: Academic Press.

HARRIGAN, W. F. & MCCANCE, M. E. (1966). *Laboratory Methods in Microbiology*. London and New York: Academic Press.

HORNSEY, H. C. & MALLOWS, J. H. (1954). Beef-curing brines. I. Bacterial and chemical changes occurring in rapidly developing, short-life brines. *J. Sci. Fd Agric.*, **5,** 573.

HUGH, R. & LEIFSON, E. (1953). The taxonomic significance of fermentative versus oxidative metabolism of carbohydrates by various Gram-negative bacteria. *J. Bact.*, **66,** 24.

INGRAM, M. (1957). Micro-organisms resisting high concentrations of sugars or salts. In *Microbial Ecology*, p. 90 (R. E. O. Williams and C. C. Spicer, eds). Cambridge: University Press.

INGRAM, M. (1958). The general microbiology of bacon curing brines, with special reference to methods of examination. In *The Microbiology of Fish and Meat Curing Brines*, p. 121 (B. P. Eddy, ed). London: HMSO.

INGRAM, M., KITCHELL, A. G. & INGRAM, G. C. (1958). Sensitive halophilic bacteria of bacon curing brines. In *The Microbiology of Fish and Meat Curing Brines*, p. 205 (B. P. Eddy, ed). London: HMSO.

JAYNE-WILLIAMS, D. J. (1963). Report of a discussion on the effect of the diluent on the recovery of bacteria. *J. appl. Bact.*, **26**, 398.

KEDDIE, R. M. (1951). The enumeration of lactobacilli on grass and in silage. *Proc. Soc. appl. Bact.*, **14**, 157.

KITCHELL, A. G. (1958). The micrococci of pork and bacon and of bacon brines. In *The Microbiology of Fish and Meat Curing Brines*, p. 191 (B. P. Eddy, ed). London: HMSO.

KOVACS, N. (1956). Identification of *Pseudomonas pyocyanea* by the oxidase reaction. *Nature, Lond.*, **178**, 703.

KUSHNER, D. J. (1968). Halophilic bacteria. *Adv. appl. Microbiol.*, **10**, 73.

LARSEN, H. (1962). Halophilism. In *The Bacteria*, Vol. IV, p. 297 (I. C. Gunsalus and R. Y. Stanier, eds). New York and London: Academic Press.

DE MAN, J. C., ROGOSA, M. & SHARPE, M. E. (1960). A medium for the cultivation of lactobacilli. *J. appl. Bact.*, **23**, 130.

MILES, A. A. & MISRA, S. S. (1938). The estimation of the bactericidal power of the blood. *J. Hyg., Camb.*, **38**, 732.

MRAK, E. M. & PHAFF, H. J. (1948). Yeasts. *Annu. Rev. Microbiol.*, **2**, 1.

PATTERSON, J. T. (1966). Characteristics of staphylococci and micrococci isolated in a bacon curing factory. *J. appl. Bact.*, **29**, 461.

PATTERSON, J. T. & CASSELLS, J. A. (1963). An examination of the value of adding peptone to diluents used in the bacteriological testing of bacon curing brines. *J. appl. Bact.*, **26**, 493.

RHODES, M. E. (1958). The cytology of *Pseudomonas* spp. as revealed by a silver-plating method. *J. gen. Microbiol.*, **18**, 639.

ROGOSA, M., MITCHELL, J. A. & WISEMAN, R. F. (1951). A selective medium for the isolation of oral and faecal lactobacilli. *J. Bact.*, **62**, 132.

SHARPE, M. E., FRYER, T. F. & SMITH, D. G. (1966). Identification of the lactic acid bacteria. In *Identification Methods for Microbiologists*, Part A, p. 65 (B. M. Gibbs and F. A. Skinner, eds). London and New York: Academic Press.

SHAW, C., STITT, J. M. & COWAN, S. T. (1951). Staphylococci and their classification. *J. gen. Microbiol.*, **5**, 1010.

SPENCER, R. (1967). *A study of the factors affecting the quality and shelf life of vacuum packaged bacon and of the behaviour of Wiltshire cured bacon packed and stored under controlled conditions.* Research Report No. 136. Leatherhead, Surrey: BFMIRA.

TOFTE JESPERSEN, N. J. & RIEMANN, H. (1958). The numbers of salt-tolerant bacteria in curing brine and on bacon. In *The Microbiology of Fish and Meat Curing Brines*, p. 177 (B. P. Eddy, ed). London: HMSO.

VARNAM, A. H. & GRAINGER, J. M. (1970). Acetate tolerance of lactobacilli from bacon curing brines. *J. appl. Bact.*, **33**, iii.

VARNAM, A. H. & GRAINGER, J. M. (1972). Enumeration of certain lactic acid bacteria from Wiltshire bacon curing brines. *J. Sci. Fd Agric.*, **23**, 546.

Microbiological Sampling in Abattoirs

A. G. KITCHELL, G. C. INGRAM AND W. R. HUDSON

Meat Research Institute, Langford, Bristol BS18 7DY, England

Bacteriological sampling within abattoirs presents problems: the slaughterhouse environment is inimical to refined or complicated sampling techniques and, as has long been recognized (Haines, 1933), many different surfaces and materials may require examination; for example, tiled walls, concrete floors, wooden or stainless steel surfaces, equipment, the hands and clothing of personnel, air, water, drains, as well as materials of animal origin, including deep muscle tissue, surfaces of meat and offals, blood and faeces.

The nature of the material will broadly determine the sampling technique employed, but other factors will influence the detailed procedure. Taking samples in an abattoir for subsequent bacteriological examination demands methods that are direct and simple. A modern abattoir may slaughter and eviscerate 120 sheep per hour, allowing little time for "on-line" sampling of each carcass moving suspended, along an overhead rail, and any but rapid (and preferably non-destructive) methods would interfere with production. This is unacceptable to management and labour alike, and creates resistance to laboratory staff and their work. Some traditional laboratory procedures are inappropriate for other reasons. For example, it is dangerous to carry about swabs in glass tubes on the slippery floors of slaughterhouses (or kitchens—*pace* Adams, James and Mazurek, 1964)—to which lacerations on the hands of one of us is ample testimony.

Should the services of a laboratory not be available locally, procedures have to be modified accordingly. For example we describe a simple loop dilution technique that reduces the amount of equipment to be carried in field studies. It is used together with pre-poured plastic Petri dishes which, failing all else, can be incubated at ambient temperature in plastic bags.

Existing Methods

Non-destructive sampling

Swabbing

Moist absorbent cotton wool swabs were used to sample areas on turkeys defined with a metal template (Walker and Ayres, 1959) and May (1961) proposed the substitution of disposable paper templates. To standardize the swabbing procedure, Reuter (1963) described an auxiliary device (trigger) by which the swabbed area and pressure applied were controlled.

The composition of the swab is important. Soluble alginate swabs are often recommended, but Patterson (1968a) reported that only one site out of seven on cattle carcasses yielded higher counts when they were used instead of cotton-gauze. Barnes (1952), sampling glass surfaces, found that calcium alginate was no more effective or reliable than gauze or absorbent cotton wool. However in 94·4% of samples of dairy equipment calcium alginate wool gave higher counts than ribbon gauze (Tredinnick and Tucker, 1951). A total body swabbing technique using calcium alginate swabs to sample successive areas of 150 cm² of poultry surface was described by Mossel and Büchli (1964). Patterson (1968b) also swabbed the entire outer surface of lamb carcasses with three small cotton gauze swabs.

A problem with all but total body swabbing techniques is the need for adequate replication because the distribution of microorganisms on meat and other surfaces is far from uniform (Empey and Scott, 1939). A rapid swabbing method taking account of this fact was devised by Hansen (1962) for use in bacon factories.

Rinsing

Most rinsing methods have limitations which restrict their application. Given an enclosed space, e.g. the diverticulum of the anterior thoracic air sac of chickens, Tarver, May and Boyd (1962) satisfactorily flushed with saline using a needle and syringe. Angelotti, Foter, Busch and Lewis (1958), while evaluating methods for determining the bacterial contamination of surfaces, scrubbed the rinse solution on the test site with rubber policemen. A constant pressure spray gun was developed by Clark (1965) for sampling poultry skin (Fig. 1); it was effective only on near-vertical surfaces. A simpler method, but restricted to near-horizontal surfaces, was devised by Williams (1967) and requires only an A1 (10 oz) can body—see p. 67.

Contact methods

Agar can be poured on non-porous surfaces and incubated *in situ* (Guiteras, Flett and Shapiro, 1954; Angelotti and Foter, 1958) or can be stripped off and incubated in a Petri dish (Hammer and Olson, 1931). Alternatively, suitable stiff agar gels can be applied to surfaces to be sampled, most conveniently in the form of "sausages" (ten Cate, 1965), and removed in slices for incubation. Such methods all suffer from the disadvantage of being inapplicable when the count is above 100 cells (or colonies)/cm^2 because the agar surface becomes overcrowded. They are excellent for the routine checking of smooth dry surfaces such as butcher's blocks or wall tiles and the results can, with advantage, be plotted according to Hansen's (1962) method (Baltzer and Wilson, 1965). They are, however, of little value for carcass meat and poultry (see e.g. Büchli, 1965).

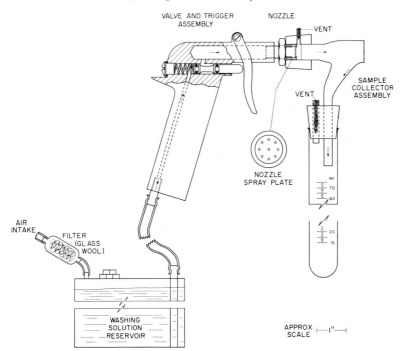

FIG. 1. The constant pressure spray gun devised by Clark (1965) for sampling poultry skin. (Reproduced by permission of the National Research Council of Canada from the *Canad. J. Microbiol.*, **11**, 1021–1022 (1965).)

A modification of the replica plating technique utilized a velvet pad mounted on a cylindrical support to transfer organisms from the surface of the sample to the surface of an agar plate (Greene, Vesley and Keenan, 1962), but recoveries were low.

The use of self-adhesive "cellophane" tape for collecting surface organisms, which were then stained and examined microscopically, was reviewed by Hartman (1968) who remarked that no one had thought of using the method for the enumeration of microorganisms. In fact, Thomas (1961) described the detection and estimation of skin staphylococci by means of such sticky tape. After application to the skin, the tape was returned to the backing paper for transport. In the laboratory, the adhesive surface of the tape carrying microbes from the sample was brought in contact with the surface of an agar plate which was then incubated. Although primarily intended as a qualitative procedure Thomas (1961) pointed out that roughly quantitative results could be obtained. Indeed, the cleaning of wooden surfaces in meat and poultry plants was monitored by this technique (Mossel, Kampelmacher and van Noorle Jansen, 1966). Counts compared favourably with those obtained by the agar sausage method but were lower than swab counts. Also of interest is the slide "profile" technique (Thomas, 1966), developed out of experience with sticky tapes, which was used in a comparison of the rates of penetration by surface bacteria into normal and tenderized beef. After surface sterilization, an incision was made to a depth of 7 cm in large cuts of beef. Flamed glass slides (7·5 × 2·5 cm) were pressed on the freshly cut surface and either stained or used to make an "impression plate" for incubation.

Moistened membrane filters were used to sample meat surfaces (Silliker, Andrews and Murphy, 1957), but the bacteria were removed by shaking the filters and were counted in the usual way after dilution. Though as reliable as a core-sampling technique, this method gave lower counts.

In addition to the comparison of methods, including contact procedures, made by Mossel *et al.* (1966), Coretti (1966) also used rinsing, swabbing, "the trigger" (Reuter, 1963) and an agar impression method to determine the state of hygiene in meat plants.

Destructive methods

Meat and offals

Haines (1937) removed a known area of superficial tissue by pressing a sterile cork borer into the meat, leaving a cylinder of tissue from the top of which a disc *c.* 2 mm thick was cut with a scalpel. The same method was also used to remove deep muscles samples after the surface was sterilized (Eddy, Gatherum and Kitchell, 1960).

Areas of 10 cm² delineated by a template were excised from the surface of livers in a recent study (Gardner, 1971). Using a template, soft materials such as meat have been sampled by scraping with a scalpel (Kitchell and Ingram, 1965). Spencer (1959) examined commercial wooden fish boxes by direct swabbing or scraping with a razor blade, wiping the scrapings onto a soluble alginate swab; counts were 2–10 times higher on the scraped samples. A triangular paint scraper used with a template has been found useful for sampling the surface of blocks of frozen meat, though the more usual method, especially for deep muscle samples, employs a brace (or electric drill) and bit (Schneiter, 1939); a useful modification for collecting the drillings has been suggested by Adams and Busta (1970)—see p. 70.

Faeces

Faecal material, e.g. in lairages, is readily collected in plastic bags. The bag is inverted over the hand to form a glove, the faecal material picked up, and the top of the bag pulled back and secured.

Should samples be required directly from the animal without contamination by dropping, a harness has been described to which plastic bags can be attached and retrieved with ease (Thomas, 1970).

Drains

The use of gauze swabs (or sanitary towels – J. H. McCoy, pers. comm.) suspended in drains for investigating salmonellosis in food premises was suggested by Moore, Perry and Chard (1952). They have since been used effectively in surveys of abattoirs (e.g. Harvey and Phillips, 1961).

Alternatively, a quantity of fluid can be withdrawn from the drain, filtered through cotton wool, and the cotton wool plug enriched for salmonellae (McCoy, 1962).

Water

The methods for sampling water are those in general use (see Report, 1969). Data were gathered in slaughterhouses by, for example, Haines (1933) and Empey and Scott (1939). Whereas mains water, of course, was satisfactory, that taken from containers used to store water for carcass washing had very high counts. It can readily be shown that water in hoses left unused over the weekend may develop a high count; such water should be run off before the hose is used for washing carcasses.

Air sampling

Haines (1933) measured aerial contamination in small and large slaughterhouses by filtration either through glass wool or a 1% sucrose solution. To determine the rate of deposition of bacteria from the air onto carcasses he exposed agar plates horizontally for 15 min, *c.* 3 ft from the floor; during the early stages of exposure, there was a linear relationship between count and exposure time. Haines found maximum rates of deposition of 48 organisms/cm^2/h and Empey and Scott (1939) 2030, though their mean value was 148 organisms/cm^2/h. Air counts in slaughterhalls ranged from 2–18 organisms/ft^3 (Haines, 1933).

More bizarre methods have been suggested, e.g. holding aloft a moistened throat swab for a time equal to that taken to swab a food contact surface (Adams *et al.*, 1964); recovery of organisms "indicated environmental atmospheric conditions existing in the food processing area"! More usual nowadays is the use of a slit sampler (Bourdillon *et al.*, 1948).

Recommended Sampling Methods

Non-destructive sampling

Template and swab

The preferred means of non-destructive sampling is the swab, though the virtues of the agar sausage for routine examination of relatively clean surfaces are recognized. A range of templates is used, mainly 50 or 100 cm^2, cut from sheet aluminium or stainless steel, which are sterilized for re-use by flaming with alcohol; disposable templates are made from thin wire.

Cotton wool balls retain their integrity better than swabs cut from rolls of cotton wool and they can be bought, ready for use, from medical suppliers. They are carried in plastic bags and handled with forceps. Gauze squares $3 \times 3 \times 1$ cm are used when a material more abrasive than cotton wool is called for, e.g. for surfaces covered with hard fat. A swab is moistened before use with sterile saline containing 0·1 % (w/v) peptone (Kelch and Friess (1959) as adopted by ISO), and rubbed first down and then across the sampling site. A second dry swab is rubbed over the site in the same manner. Both swabs are collected in a wide-necked 1 oz bottle containing 10 ml of diluent. These tubes are more robust than test tubes and are carried in a stout, partitioned cardboard box. Three sites per meat unit (side or packed joint) are swabbed and treated separately; where the number of units to be sampled is too large, the swabs from 3 sites are bulked.

The 1 oz bottles containing both swabs in diluent are shaken on a vortex mixer (Fisons "Whirlimixer") before dilution and plating (see p. 54).

Examples of use

The scatter of counts from different 50 or 100 cm^2 areas of the exterior and interior of cattle, sheep and pig* carcasses is illustrated in Table 1. The same method applied to vacuum packed primal cuts of beef (3 samples of 100 cm^2 from the top, side and bottom surfaces) demonstrated the influence of the age in pack on the count (Table 2). Monitoring of a new slaughter facility by swabbing established a slow build-up of bacteria on the floor (Fig. 2). Thorough cleaning reduced counts by a factor of 100.

TABLE 1. The mean values and range of counts for swabs of various sites on six lamb, cattle and pig carcasses

	Log viable count/cm^2 at 20°	
Lamb (10 sites)	Cattle (13 sites)	Pig (12 sites)*
3·81	2·86	4·02
(3·08–4·19)	(1·98–3·78)	(2·74–6·43)
3·72	2·89	5·02
(2·47–4·42)	(1·51–3·88)	(4·16–5·84)
3·22	3·52	4·50
(2·58–4·36)	(2·21–4·80)	(3·66–5·06)
3·13	2·96	4·07
(2·42–3·47)	(2·36–3·80)	(3·58–4·44)
3·32	2·73	4·75
(2·19–5·04)	(1·21–5·20)	(3·53–5·43)
3·27	2·64	4·77
(2·18–4·42)	(1·04–4·20)	(3·13–5·36)

TABLE 2. Relation of counts on vacuum packed tenderized beef to the age in pack

Days in vacuum pack	No. samples (100 cm^2)	Log viable counts/cm^2 incubated at 37°	25°	1°
6	18	3·59	4·28	4·18
7	3	4·31	4·44	4·46
13	12	4·29	5·03	4·99
15	3	4·70	6·31	6·35
20	3	6·58	6·81	6·71
21	3	6·50	7·03	7·16

* The counts quoted are high for pigs in general but are typical of the premises sampled, where the mean count at 20° for 80 pig carcasses was 4·47 (range 2·29–5·82).

Total body swab

The examples quoted above were of unhurried sampling during convenient breaks in slaughtering or in butchers' cutting rooms; the template technique is less convenient for use on carcasses on slaughterlines, especially by one person. For that purpose, a simple total body swabbing technique was developed. It permits sampling with one hand whilst the other steadies the swinging carcass.

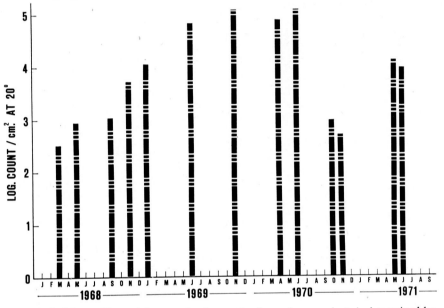

FIG. 2. Build-up of bacterial contamination on the floor of a new abattoir determined by swab and template (100 cm²) and the effect of cleaning up (autumn, 1970).

Dr White's No. 1 sanitary towels* with loops removed are autoclaved in foil at 121° for 15 min and then dried in a plate-drying oven at 50° for c. 3 h. The dry sterile towels are transferred aseptically into individual polythene bags (c. 20 × 30 cm) which, as purchased, are found to be virtually sterile and the bags are sealed with rubber bands. When required, two swabs are taken and 50 ml of sterile diluent (saline containing (w/v) 0·1% peptone) is added to one bag.

In use, the wet swab is grasped through the bottom of the outside of the bag and the upper part of the bag pulled back over the hand forming a protective glove. The exposed swab, is rubbed over the exterior and interior surfaces of the carcass. It is then covered immediately by folding back the

* An assurance from the manufacturer that no anti-microbial agents are incorporated in the sanitary towels, has been confirmed by tests on towels in use as swabs.

polythene bag. The dry swab is now used in the same manner and put with the wet swab. The plastic bag in which the dry swab was carried is used to form a double envelope for protection of both swabs during transport to the laboratory. The time required varies according to size of the carcass, e.g. less than 2 min for lamb, but is always less than that required to swab 3 areas with a template, though physically more demanding.

In the laboratory, a further 250 ml of saline/peptone diluent are added to the swabs in the polythene bag which are kneaded by hand for *c*. 30 sec to wash off bacterial cells into suspension. An Astell pipette (1·0 ml) is introduced directly into the bag to withdraw an aliquot for dilution. In the field, a loop would be used (see p. 57).

In many practical situations comparative results only are required and the fact that results obtained by this method cannot be expressed as is usual, per unit area, but only per carcass, is not a serious disadvantage. In any case, estimates can be made of the surface area of carcasses e.g. 10^4 cm^2 for lamb, and the counts converted for comparison with data expressed as counts/cm^2. The method avoids the problems of sampling arising from the variation in count at different points on the carcass (Table 1) and has a decisive advantage if the purpose is to detect organisms such as salmonellae that may not be widely distributed over the surface.

Examples of use

The method has been used to evaluate the spray washing of carcasses, both "on-line" in abattoirs and in the Meat Research Institute slaughterhouse. In the abattoirs, lambs were examined, half of the surface area per carcass (i.e. one side) being sampled before the final spray was applied and half afterwards; the effects of the spraying ranged from a 90% reduction in bacterial count to a 25% increase (Fig. 3).

The method has also been applied successfully in surveying the incidence of salmonella on carcass meat. The swabs are best directly enriched but if a total count is also required the swab washings can be filtered (see p. 56) and added to an enrichment medium together with the swabs.

Direct comparison of template and total body swabbing is difficult for practical reasons. Tests made on adjacent 100 cm^2 areas of carcass using cotton wool balls and sanitary towels followed by excision of the areas, maceration and counting, indicate the superiority of the large swab. The cotton wool recovered 25% of the total number of bacteria on the site whereas the sanitary towel removed 58%.

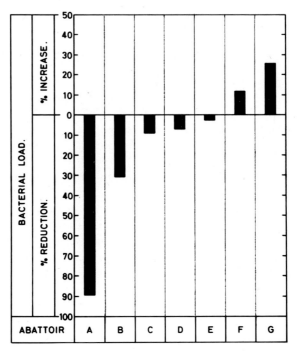

F<small>IG</small>. 3. Comparison of carcass cleaning methods used in commercial abattoirs: all used water sprays, A with hot water, B and C with warm water, D and G with cold water (in G, wiping cloths were also used), E with cold water and a scrubbing brush, and F with warm water and a car wash brush. Results based on left and right sides of 12 carcasses in each abattoir.

Destructive sampling

Meat

Weighed samples. In many instances, samples taken for other purposes (chemical composition, pesticide residues, etc.) are also examined bacteriologically. The chemical tests require weighed samples and destructive sampling techniques are used. An important consideration here is the commercial value of e.g. a side of beef which, if drilled with holes or otherwise disfigured, would no longer be saleable. Sampling in these circumstances, therefore, must do minimum damage to the meat. Our proposal, recently adopted by the International Organization for Standardization (ISO) is that samples be taken from the neck area of carcasses—neck muscles from cattle, several vertebrae from sheep and the head of the pig. These tissues yield sufficient material for examination (Table 3) and the sites are frequently those giving the highest count on beef carcasses (Scott and Vickery, 1939) and on pigs (Table 4).

TABLE 3. Yields of tissue, from the neck end of carcasses of normal slaughter weight, suitable for bacteriological examination

| Tissue | Weight (g) of tissue from | | |
	Lamb (neck)	Pork (cheek and neck muscle)	Beef (neck muscle)
Muscle	401	464	669
Fat	249	178	581
Bone	374	—	—
Rind	—	256	—
Total wt (g)	1024	898	1250

TABLE 4. A comparison of counts on 50 g samples of different tissues from a pig carcass

| Site | Log viable count/g incubated at | | | Geometric mean |
	37°	20°	1°	
Cheek	6·8	8·3	8·3	7·81
Shoulder	6·2	7·9	7·9	7·34
Ham	5·6	7·9	8·0	7·18
Loin	5·8	8·2	8·2	7·41

Core samples. (a) Counts on a meat surface can be made most accurately by removing the surface tissues with a cork borer (Haines, 1937) and, where the cost of the meat need not be considered, this is the best method of sampling.

(b) The same technique permits sampling of deep muscle tissue, for example, in the investigation of bone taint. Here, however, it is essential to avoid contamination of the core sample with bacteria from the surface. Therefore, a large diameter cork borer and scalpel are used to remove the surface tissue to a depth of approximately 4 mm. The exposed area is then seared with a hot iron of the same diameter and a smaller diameter borer is introduced at the centre to obtain a deep muscle sample (Eddy *et al.*, 1960).

Preparation for plating. Following recommendations made by ISO on the basis of a collaborative study (Barraud *et al.*, 1967) weighed samples are put twice through a sterile mincer with a cutting plate having holes of *c.* 4 mm diam. A weighed sample of the mince is then macerated with 9 volumes of diluent pre-chilled to 1° in a mechanical blender for 15,000–20,000 revolutions. The results obtained in the ISO tests with different methods of maceration (Table 5) show that any mechanical method is better than a manual one (in this instance, grinding with sand), especially if the meat is tough.

The object of specifying the number of cutting revolutions, apart from standardizing the procedure, is to limit the time of operation of mechanical blenders, otherwise unwanted temperature rises can be recorded. For example, starting with 20 g of meat in 200 ml of diluent at 22° the temperature after 5 min may be as high as 42° in a 500 ml blender cup (MSE Atomix). Therefore, we recommend the use of chilled diluent for maceration.

TABLE 5. A comparison of the counts obtained at 30° on samples of tough and tender beef macerated in different ways (data of Barraud et al., 1967)

Method of maceration		Viable counts (%)* on	
Apparatus	Operation (sec/rpm)	tough beef	tender beef
MSE Atomix	30/6000 +60/12,000	100	100
Biorex	30/45,000	94	96
Ultra-Turrax	30/23,000	95	93
MSE Nelco	120/8000	94	106
Pestle and mortar (with sand)	120 sec	55	103

* Mean counts by all other methods expressed as percentages of the Atomix count.

Core samples taken with a cork borer may be treated in the same way, though mechanical shaking under standardized conditions (see Gardner and Kitchell, p. 15) is equally efficacious. Indeed, earlier work (Kitchell in Jayne-Williams, 1963) established that higher counts were consistently obtained by shaking with glass beads as compared with maceration in an MSE Atomix blender.

Decimal dilutions are made in saline + 0·1% peptone. Usually, 1/50 ml drops of selected dilutions are transferred in duplicate to the surface of poured and dried plates of Standard Methods Agar (APHA, 1960) e.g. Plate Count agar (Oxoid), by means of calibrated dropping pipettes (see Gardner and Kitchell, p. 16). In general, drops of 4 successive dilutions are spread over $\frac{1}{4}$ plate with a glass spreader or platinum loop but 0·1 ml can be spread over a whole plate, as recommended by ISO. Incubation is usually carried out at 37, 25–20, and 1°. If a single temperature has to be selected, 30° is to be preferred. A rapid dilution and plating method devised for field studies is described on p. 57.

Wiping cloths, brushes, etc.

Although now proscribed by law, the mutton cloth wiper used to finish off the dressing of carcasses may be replaced, in current practice, with a woven paper disposable tissue or a paper towel. The method of sampling is the same in all cases. An area is cut from the wiping material when it is discarded by the slaughterman and is macerated mechanically. The build-up of bacteria on a wiping cloth used on successive carcasses and then incubated for 8 h (Fig. 4) illustrates why its use is no longer permitted. In abattoir G (Fig. 2) wiping cloths were still being used.

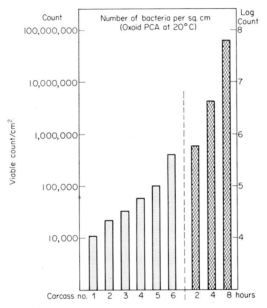

FIG. 4. Bacterial counts (at 20°) on a wiping cloth used on 6 successive beef carcasses then incubated for 8 h.

However, in its place we find all manner of spraying devices, with and without brush attachments. The latter can contribute to carcass contamination—the opposite effect to that intended (Fig. 2, F)—as was established by removing bristles, shaking them with beads, and counting. Counts were as high as $3·45 \times 10^7$ organisms/bristle.

Carcass wash water

In surveys of salmonellae in slaughterhouses, direct examination of the water used in spray washing of carcasses may be of value. The water draining from a carcass is collected, with or without the aid of a large diameter funnel, in a sterile bucket. It is then filtered (McCoy, 1962) and the filter enriched in the usual way for salmonella. From a total of 200 pigs so far examined, *Salmonella typhimurium* was isolated 7 times, *S. albany* twice. and *S. heidelberg* and *S. panama* once.

Air samples

Three air samplers are used by us (Fig. 5). Two are slit samplers (Casella, Mk 11, Britannia Walk, London, N1; Reynier, Chicago 13, Ill., USA) in which the air impinges on agar plates and the third draws air through a

FIG. 5. Three air samplers: Fisons, top left; Reynier, top right; and Casella, bottom.

membrane filter (Fisons, Loughborough, England). The Casella is the most flexible in use, the most efficient (see Table 6) and the heaviest. The Reynier is more readily transported than the Casella, but the smaller Fisons instrument is most convenient for field studies. Typical results in three abattoirs are given in Table 7. The efficacy of fumigation and cleaning procedures in the chilled meat lockers of ships has also been monitored many times with air samplers; formaldehyde treatment is shown to reduce aerial contamination by $c.\ 90\%$.

TABLE 6. Relative efficiency of three air samplers

Sampler	Type	%†	Viable count* /ft³
Casella	Slit	100	66
Fisons	Filter	62	41
Reynier	Slit	58	38

* Geometric means of 20 samples.
† Casella count taken as 100 %.

TABLE 7. Typical air counts in three abattoirs and a meat preparation room

Site	Abattoir	Viable count/ft³
Lairage	A	452
Slaughterhall	A	17
	B	20
	C	44
Chill room	A	6
	B	8
	C	3
Meat pre-packing room	D	77

Rapid Dilution and Plating Method

At the same time as the large swab was developed for survey work in abattoirs, a rapid method of diluting and plating samples requiring the minimum of apparatus was also devised. The basis is the loop dilution and inoculation method practised in dairy bacteriology for more than 50 years. Only a calibrated platinum wire loop delivering 0·02 ml drops, a glazed spotting tile, and pre-poured agar plates (plastic) are required. Nine drops of sterile diluent are transferred to the depressions of the tile with the loop. A loopful of the swab washings or tissue macerate is transferred to the first depression and mixed with diluent. The loop is sterilized with a portable butane torch (e.g. Flamidor, Southern Watch & Clocks Supplies Ltd., Orpington, Kent), cooled, and used to make the next dilution or, if the first dilution is to be plated, a drop is first transferred to the surface of a plate and spread over one quarter. Dilution and plating proceed, with sterilization of the loop between each step, until the appropriate range of dilutions has been prepared and plated.

The method gives counts comparable with those made in a more orthodox manner (Fig. 6) and this was confirmed by G. A. Gardner (pers. comm.)

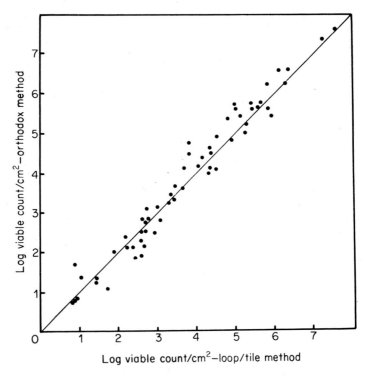

FIG. 6. Comparison of counts made by an orthodox dilution and plating method and by the loop/tile technique. Twenty sites in an abattoir were swabbed and the data plotted are the counts at 37, 20, and 1°.

in a survey of bacon in several factories. He reported also that samples were handled in $\frac{1}{3}$–$\frac{1}{4}$ of the time required by traditional methods. In field work, if a laboratory is not available, the procedure is simple enough to be carried out on any table and the plates can be incubated, loosely packed in plastic bags, at ambient temperature. Armed with the large swabs and the rapid culture technique a survey of hygiene practices in abattoirs is about to be undertaken.

Acknowledgements

Figs 3 and 4 are reproduced by kind permission of the Institute of Meat in whose Bulletin they first were published.

References

ADAMS, D. M. & BUSTA, F. F. (1970). Simple method for collection of samples from a frozen food. *Appl. Microbiol.*, **19**, 878.

ADAMS, R. A., JAMES, W. R. & MAZUREK, E. (1964). A report on the use of swabs for improving compliance with food ordinance requirements. *Q. Bull. Ass. Fd Drug Off.*, **28**, 126.

AMERICAN PUBLIC HEALTH ASSOCIATION (1960). *Standard Methods for the Examination of Dairy Products.* 11th Ed. New York: APHA Inc.

ANGELOTTI, R. & FOTER, M. J. (1958). A direct surface agar plate laboratory method for quantitatively detecting bacterial contamination on nonporous surfaces. *Fd Res.*, **23**, 170.

ANGELOTTI, R., FOTER, M. J., BUSCH, K. A. & LEWIS, K. H. (1958). A comparative evaluation of methods for determining the bacterial contamination of surfaces. *Fd Res.*, **23**, 175.

BALTZER, J. & WILSON, D. C. (1965). The occurrence of clostridia on bacon slaughter lines. *J. appl. Bact.*, **28**, 119.

BARNES, J. M. (1952). The removal of bacteria from glass surfaces with calcium alginate, gauze and absorbent cotton wool swabs. *Proc. Soc. appl. Bact.*, **15**, 34.

BARRAUD, C., KITCHELL, A. G., LABOTS, G., REUTER, G. & SIMONSEN, B. (1967). Standardization of the total aerobic count of bacteria in meat and meat products. *Fleischwirtschaft*, **47**, 1317.

BOURDILLON, R. B., LIDWELL, O. M. & LOVELOCK, J. E. with CAWSTON, W. C., COLEBROOK, L., ELLIS, F. P., VAN DEN ENDE, M., GLOVER, R. E., MACFARLAN, A. M., MILES, A. A., RAYMOND, W. F., SCHUSTER, E. & THOMAS, J. C. (1948). Studies in air hygiene. *Spec. Rep. Ser. med. Res. Coun. No. 262*. London: HMSO.

BÜCHLI, K. (1965). Microbiological control in poultry processing plants with the "ten Cate sausage". *Bull. int. Inst. Refrig.* **45**, Sect. 3., 227.

TEN CATE, L. (1965). A note on a simple and rapid method of bacteriological sampling by means of agar sausages. *J. appl. Bact.*, **28**, 221.

CLARK, D. S. (1965). Method of estimating the bacterial population on surfaces. *Can. J. Microbiol.*, **11**, 407.

CORETTI, K. (1966). The value of some bacteriological methods of determining the state of hygiene in meat factories. *Fleischwirtschaft*, **46**, 139.

EDDY, B. P., GATHERUM, D. P. & KITCHELL, A. G. (1960). Bacterial metabolism of nitrate and nitrite in maturing bacon. *J. Sci. Fd Agric.*, **11**, 727.

EMPEY, W. A. & SCOTT, W. J. (1939). Investigations on chilled beef. Part I. Microbial contamination acquired in the meat works. *CSIRO, Bull. No.* **126**. Melbourne: CSIRO.

GARDNER, G. A. (1971). A note on the aerobic microflora of fresh and frozen porcine liver stored at 5°C. *J. Fd Technol.*, **6**, 225.

GREENE, V. W., VESLEY, D. & KEENAN, K. M. (1962). New method for microbiological sampling of surfaces. *J. Bact.*, **84**, 188.

GUITERAS, A. F., FLETT, L. H. & SHAPIRO, R. L. (1954). A quantitative method for determining the bacterial contamination of dishes. *Appl. Microbiol.*, **2**, 100.

HAINES, R. B. (1933). Observations on the bacterial flora of some slaughterhouses. *J. Hyg., Camb.*, **33**, 165.

HAINES, R. B. (1937). Microbiology in the preservation of animal tissues. *DSIR Fd Invest. Spec. Rep. No.* **45**. London: HMSO.

HAMMER, B. W. & OLSON, H. C. (1931). Bacteriology of butter. 3. A method for studying the contamination from churns. *Res. Bull. Iowa agric. Exp. Stn No.* **141.**

HANSEN, N.-H. (1962). A simplified method for the measurement of bacterial surface contamination in food plants and its use in the evaluation of pressure cleaners. *J. appl. Bact.*, **25**, 46.

HARTMAN, P. A. (1968). Cellophane tape for collection (and culture?) of specimens. *Adv. Appl. Microbiol. Suppl.* **1**, 57. In *Miniaturized Microbiological Methods*. New York and London: Academic Press.

HARVEY, R. W. S. & PHILLIPS, W. P. (1961). An environmental survey of bakehouses and abattoirs for salmonellae. *J. Hyg., Camb.*, **59**, 93.

JAYNE-WILLIAMS, D. J. (1963). Report of a discussion on the effect of the diluent on the recovery of bacteria. *J. appl. Bact.*, **26**, 398.

KELCH, F. & FRIESS, H. (1959). Investigations on the performance of bacteriological control in meat processing plants. *Fleischwirtschaft.*, **11**, 1011.

KITCHELL, A. G. & INGRAM, M. (1965). The effect on the bacterial flora of Wiltshire bacon of feeding sugar to pigs before slaughter. *Proc. 1st Int. Cong. Fd Sci. Technol., London*, **2**, 105. London: Gordon & Breach.

MAY, K. N. (1961). Skin contamination of broilers during commercial eviceration. *Poult. Sci.*, **40**, 531.

McCOY, J. H. (1962). The isolation of salmonellae. *J. appl. Bact.*, **25**, 213.

MOORE, B., PERRY, E. L. & CHARD, S. T. (1952). Survey by sewage swab method of latent enteric infection in an urban area. *J. Hyg., Camb.*, **50**, 137.

MOSSEL, D. A. A. & BÜCHLI, K. (1964). The total object swab ("TOS") technique. A reliable and convenient method for the examination of some proteinaceous staple foods for various types of enterobacteriaceae. *Lab. Pract.*, **13**, 1184.

MOSSEL, D. A. A., KAMPELMACHER, E. H. & NOORLE JANSEN, L. M. VAN. (1966). Verification of adequate sanitation of wooden surfaces used in meat and poultry processing. *Zentbl. Bakt. ParasitKde Bakt. 1. Orig.*, **201**, 91.

PATTERSON, J. T. (1968a). Hygiene in meat processing plants. 2. Methods of assessing carcass contamination. *Rec. Agric. Res. (Min. Agric. N. Ireland)*, **17, (1)**, 1.

PATTERSON, J. T. (1968b). Hygiene in meat processing plants. 3. Methods of reducing carcass contamination. *Rec. Agric. Res. (Min. Agric. N. Ireland)*, **17, (1)**, 7.

REPORT (1969). The bacteriological examination of water supplies. *Dep. Hlth Soc. Sec., Rep. No.* **71**. London: HMSO.

REUTER, H. (1963). Suggestion for standardized sampling in the bacteriological plant control. *Fleischwirtschaft*, **15**, 195.

SCHNEITER, R. (1939). Report of microbiological methods for examination of frozen egg products. *J. Ass. off. agric. Chem.*, **22**, 625.

SCOTT, W. J. & VICKERY, J. R. (1939). Investigations on chilled beef. 2. Cooling and storage in the meat works. *CSIRO, Bull. No.* **129.** Melbourne: CSIRO.

SILLIKER, J. H., ANDREWS, H. P. & MURPHY, J. F. (1957). A new non-destructive method for the bacteriological sampling of meats. *Fd Technol., Champaign.*, **11**, 317.

SPENCER, R. (1959). The sanitation of fish boxes. 1. The quantitative and qualitative bacteriology of commercial wooden fish boxes. *J. appl. Bact.*, **22**, 73.

TARVER, F. R. JR., MAY, K. N. & BOYD, F. M. (1962). Sampling techniques for the enumeration of microorganisms in the diverticulum of the anterior thoracic air sac of chickens. *Appl. Microbiol.*, **10**, 137.

THOMAS, J. G. (1970). A method of faeces collection from sheep. *Lab. Pract.*, **19**, 75.

THOMAS, M. (1961). The sticky film method of detecting skin staphylococci. *Mon. Bull. Minist. Hlth*, **20**, 37.

THOMAS, M. (1966). Bacterial penetration in raw meats: comparisons using a new technique. *Mon. Bull. Minist. Hlth*, **25**, 42.

TREDINNICK, J. E. & TUCKER, J. (1951). The use of calcium alginate wool for swabbing diary equipment. *Proc. Soc. appl. Bact.*, **14**, 85.

WALKER, H. W. & AYRES, J. C. (1959). Microorganisms associated with commercially processed turkeys. *Poult. Sci.*, **38**, 1351.

WILLIAMS, M. L. B. (1967). A new method for evaluating surface contamination of raw meat. *J. appl. Bact.*, **30**, 498.

The Sampling of Chickens, Turkeys, Ducks and Game Birds

ELLA M. BARNES, C. S. IMPEY

Food Research Institute, Colney Lane, Norwich NOR 70F, England

AND

R. T. PARRY

Bernard Matthews Ltd., Great Witchingham Hall, Norwich NOR 65X, England

There is no single sampling procedure which can be applied in the microbiological examination of poultry because there are so many different ways of processing the bird. Most chickens and many of the turkeys and ducks are killed, plucked, eviscerated, air or water chilled and then either held chilled or frozen until bought by the consumer. Some birds are stored plucked, but uneviscerated. Birds such as pheasants, partridges and other game birds are generally hung at a cool ambient temperature (*c.* 10°) and only plucked and eviscerated immediately before cooking. In addition to whole carcasses, there are also a number of poultry products such as cut-up portions and various cooked turkey or chicken products to be considered.

The type of sample and the examination required will depend on the nature of the question being asked. Are there any food poisoning bacteria present? What is the general bacteriological condition and the potential storage life at a given temperature? Should one sample the area likely to be the most contaminated or is one trying to obtain an overall assessment of the condition of the carcass? These points will be discussed below in relation to the different types of product.

Chilled or Frozen Eviscerated Chickens and Turkeys

A carcass leaving a processing plant will carry many different types of bacteria, the most important being food poisoning bacteria such as salmonellae, *Clostridium welchii* or *Staphylococcus aureus*, and the psychrophilic spoilage bacteria, *Pseudomonas* and *Acinetobacter*. The origin of these organisms together with their minimum temperatures for growth is shown in Table 1. During storage under chill conditions (0–5°), little change will occur in the numbers of food poisoning bacteria but the psychrophilic organisms will multiply and cause spoilage.

Frozen carcasses which have been stored inefficiently for prolonged periods at temperatures only just below freezing (−3° to −7°) may be subject to mould spoilage. The organisms most commonly implicated (Haines, 1931) are *Cladosporium herbarum* ("black spot"), *Thamnidium elegans, Thamnidium chaetocladioides* ("whiskers") and *Sporotrichum carnis*.

Distribution of organisms on the carcass

Considerable variation can be found in the numbers and types of bacteria in different parts of the carcass. This may depend partly on variation in the processing methods used in different countries. In France a comprehensive study of chickens (Lahellec and Meurier, 1970) showed considerable variation between sites on the same carcass. On the basis of total counts at 20° for 4 days or psychrophilic counts at 3° for 3 weeks, no single site on the skin surface was consistently more contaminated than any other. They also compared skin with cavity samples and found *Escherichia coli* in considerably greater numbers in the latter. In the USA, Kotula (1966), carrying out total counts at 35°, found that the numbers of bacteria on the thighs were significantly higher than those on the breast. The present authors have also found lower total counts at 20° on the breast of turkeys when compared with the neck skin, the surface under the wing or by the side of the vent.

After storage the spoilage flora will be found growing mainly on the cut muscle surfaces and down in the holes left by the removal of the feathers. This can be demonstrated by spraying the carcass with tetrazolium solution (1% w/v) as in the method of Bradshaw, Dyett and Herschdoerfer (1961). The technique also shows that bacterial growth appears to be much less over the breast than the legs. Ziegler, Spencer and Stadelman (1954) showed that contamination was greatest in the area under the wing. When considering stored carcasses it is also possible that the types of bacteria growing in different parts of the carcass may vary. Barnes and Impey (1968) showed that pigmented and nonpigmented strains of *Pseudomonas* grew equally well in minced leg and breast muscle. Strains of *Acinetobacter* grew in leg but failed to grow in breast muscle. A spoilage organism, *Pseudomonas putrefaciens*, grew much faster in leg than in breast muscle. The differences could be explained partially by differences in the pH between these two areas.

It is unlikely that there will be any change in the distribution of bacteria of public health significance (*Staphylococcus aureus, Clostridium welchii* or salmonellae) after refrigerated storage as these bacteria do not grow at this temperature (Table 1). Care should be taken, however, in interpreting

TABLE 1. The origin of food poisoning and food spoilage organisms on poultry carcasses and their lowest temperatures for growth*

	Origin	Approximate minimum temperature for growth (°C)
Food poisoning organisms		
Staphylococcus aureus	Nasal cavity, skin, hocks, feet, etc.	10
Clostridium welchii	Intestines	15
Salmonella spp	Intestines	7
Food spoilage organisms		
Pseudomonas spp	Feathers, feet, rural water supplies,	− 2
Acinetobacter spp	general environment (not the intestines)	
Moulds and yeasts	Feathers, general environment	−5 to −7

* Data of Haines (1934), Michener and Elliott (1964) and the authors.

counts of faecal indicator organisms because some of them, e.g. the faecal streptococci, can multiply slowly at low temperatures (Foter and Rahn, 1936).

Sampling method and organisms isolated

The sampling of a frozen carcass is difficult and it is the practice of the authors to allow the carcass to thaw at 1–5° overnight before sampling. The type of sampling method used will depend on whether one is carrying out a detailed investigation of the numbers and types of bacteria present under various conditions of processing and storage or whether one is using a method for routine control and monitoring of the condition of the carcass. Once the sample has been taken the methods used for isolating the organisms tend to be the same for all foods so they will not be discussed in detail here. For a discussion of the food poisoning bacteria and faecal indicator organisms, see p. 233. If streptococci are used as faecal indicators, it should be noted that either *Streptococcus faecalis* or *Streptococcus faecium* may be found in the intestine of the bird, thus one of the media selecting for both these organisms is required.

The isolation and characterization of the poultry spoilage flora have been described (Barnes and Thornley, 1966). The majority of these organisms will not grow at 37° so that total counts at this temperature are no indication of the numbers of spoilage organisms present. Counts may be done at 20° for 3–4 days or, if a count of only the psychrophilic bacteria is required, this may be made by incubating the Petri dishes at 1° for 2 weeks.

For the isolation of moulds and yeasts, a medium such as Bacto YM Agar (Difco) pH 3.5 may be used, the plates being incubated at 20° for 5 days.

The sampling of whole skin

In a number of studies it has been shown that macerating samples of whole skin gives a higher recovery of organisms than any other technique (Fromm, 1959; Avens and Miller, 1970). This is probably due to the bacteria growing down in the feather follicles rather than at the surface of the skin.

Various types of skin samples have been used. Barnes and Shrimpton (1958) took under aseptic conditions skin (2 g) from under the wing together with some (3 g) from near the vent. The combined sample represents an area of *c*. 50 cm². Over the years this method has been used successfully by one of the authors (EMB). By sampling carcasses taken from the processing line, it was shown in the original study, that the higher count of psychrophilic bacteria obtained when the skin was macerated was due partly to *Acinetobacter* spp (then called *Achromobacter*), some of which through forming gelatinous colonies may be difficult to remove by swabbing. Thus it may not only be total numbers which differ with the type of sampling method but also the types of organisms may vary. Avens and Miller (1970) used a sterile cutting cylinder to remove a skin sample of 7·145 cm², which is possibly more accurate than always making the assumption that 1 g skin = 10 cm².

The neck flap provides a skin sample without damage to the carcass. This method has been widely advocated, particularly for the recovery of food poisoning bacteria such as salmonellae and clostridia (Simonsen, 1971; Mead and Impey, 1970). In a comparison with three other methods, Simonsen (1971) found that it gave the highest recovery of bacteria.

Sampling by skin scraping

A method described by Williams (1967) has been used with considerable success with turkey carcasses by one of the authors (R. T. Parry). This technique involves pressing a sterile, sharp-edged cylinder on to the surface of the carcass so that a known area is encircled (Fig. 1). A 10 oz (A1 size) can body can be used which gives a sampling area of *c.* 33·86 cm². Into the cylinder is poured 25 ml diluent which is then stirred with a sterile spatula and the skin surface scraped to release the maximum number of organisms. Williams (1967) considered that a tenfold increase was obtained in the numbers of bacteria recovered as compared with the swab method.

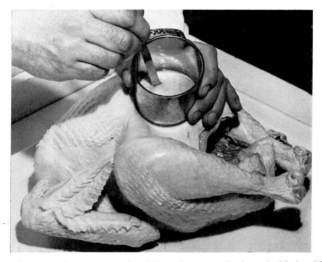

FIG. 1. Sampling by scraping the skin, using a can body to hold the diluent.

Rinsing techniques—whole carcasses

Techniques where a whole chicken is shaken in a plastic bag with *c.* 1000 ml diluent have been developed mainly in hygiene studies of freshly processed poultry in order to recover very low levels of bacteria, such as salmonellae, which are randomly distributed over the carcass. Various studies of different versions of this technique have been made and Simonsen (1971) found that rinsing a whole chicken in 1000 ml diluent by shaking vigorously for 30 sec gave a higher recovery than by dipping a carcass 15 times in 2000 ml diluent as suggested by Leistner and Szentkuti (1970).

In order to convert the number of bacteria/ml rinse water to the number/cm², it was found that the surface of a broiler chicken expressed as cm² equals approximately the weight (g) + 500 (Simonsen, 1971). An example of the successful use of this method is given in the paper of Surkiewicz, Johnston, Moran and Krumm (1969). The disadvantages of this method are evident. There is no suggestion as to how to cope with very large carcasses, particularly those of turkeys which may weigh from 2500–20,000 g. The method will only harvest those organisms which wash off easily.

Rinsing technique—defined areas

The use of a high pressure spray to rinse a defined area of carcass has been developed (Clark, 1965*a*, *b*) – see p. 45. In comparative tests using chicken skin inoculated with *Pseudomonas fluorescens*, it was shown that recoveries were superior to those obtained with conventional methods. Information is now required on whether or not this method will be as successful when used in practice where bacteria may be lodged down the feather follicles and where certain types (e.g. *Acinetobacter* spp) may not be so easily removed as are *Ps. fluorescens*.

The swab

The use of a cotton wool or alginate swab for recovering bacteria from a known area is perhaps the method most widely used for sampling surfaces. A brief description and evaluation of the various modifications of this technique have been given by Patterson (1971). In tests with poultry carcasses, the swab technique has given a poorer recovery of bacteria than the sampling methods described above.

Agar impression methods

For a discussion of these methods reference should be made to Patterson (1971). Although they generally show a lower recovery than the methods listed above, they have the great advantage of simplicity of use and equipment required. They are perhaps most effective in the monitoring of poultry processing plants for particular types of organisms. Büchli (1965) demonstrated the effective use of the "ten Cate sausage" (ten Cate, 1965) in the tracing of faecal contamination on poultry carcasses at all stages in processing.

Ducks

No specific mention has been made above of the sampling of duck carcasses because very little information has been found concerning them. However, with the exception of the hot wax de-feathering bath, the processing method is similar to that of chicken and the basic problem of distribution of organisms and sampling techniques are probably the same. Studies relating to the distribution of salmonellae in duck plants have been made by Galton,

Mackel, Lewis, Haire and Hardy (1955), Brobst, Greenberg and Gezon (1958) and Woodburn and Stadelman (1968).

Raw Products

Cut-up portions

The growing demand for poultry in the form of cut-up, chilled or frozen portions can present a microbial problem since the additional handling inherent in the processing inevitably increases the level of bacterial contamination. The numbers of spoilage bacteria are of particular importance in relation to the shelf life of the chilled product.

Non-destructive sampling methods have their limitations, the swab or template method being difficult to apply owing to the irregular surfaces to be sampled. However, Mercuri and Kotula (1963) observed a relationship between the counts obtained with the exudate or drip from packaged portions and those from swabs. Drewniak, Howe, Goresline and Bauch (1954) recommended a "shake" technique and Goresline and Haugh (1959) developed a method whereby bacterial numbers could be related to the surface area from which they came.

As discussed for whole carcasses, the majority of spoilage bacteria are found growing on the surfaces of cut muscle. Therefore a sampling method aimed at these specific areas would appear to give the best index of the shelf life and removing 5 g of tissue would not in any way reduce the acceptability of the portions. Analyses for the possible presence of food poisoning bacteria can be carried out using any of the methods discussed above.

Giblets

With the giblets (neck, gizzard, liver and heart) the greatest concern is for the detection of organisms of public health importance and particularly the general level of faecal contamination, as inevitably some contamination will have occurred during evisceration. The giblets are frequently sold in the form of a "soup" pack. They are best sampled by the removal of a weighed quantity of tissue.

Deboned poultry meat

Deboned poultry meat is widely used in the manufacture of "convenience" foods. For ease of handling and storage the raw meat is moulded into large blocks and frozen. In the course of preparation, bacteria will be distributed throughout the block, therefore, before processing, the bacterial status of each block usually needs to be assessed. Because so much handling has

D

occurred, it is particularly important to test for the presence of staphylococci and other food poisoning bacteria.

Once the block has thawed, sampling presents no problems as any of the
standard methods can be used. In the frozen state, however, obtaining a
representative sample is less simple. The traditional method in the meat
industry has been to use a brace and a sterile bit to bore out the frozen
tissue. A considerable improvement of this method was suggested by
Adams and Busta (1970) who collect their samples in a sterile funnel from
an auger-type bit fitted to a variable speed drill. The authors also favour the
use of a variable speed electric drill but have found a 5/8 in. Stanley Bridges
"plug cutter" to be preferable to the auger-type bit (Fig. 2). A sterile,
dished metal deflector is fitted between the bit and the chuck so that air
currents from the drill's cooling system do not contaminate the sample.

FIG. 2. Sampling frozen meat blocks using a ⅝ in. Stanley Bridges "plug cutter" fitted
to a variable speed electric drill.

Cooked Poultry Products

Barbecued poultry

Barbecued chickens have been implicated in a number of outbreaks of
food poisoning in recent years (Hobbs, 1971). The majority of these have
pinpointed certain inherent weaknesses in the retailing practices applied
to such products. Some of the outbreaks have been attributed to inadequate
thawing of contaminated, frozen carcasses prior to cooking, so that the
heat treatment has been insufficient to destroy the causative organisms,
whilst others have occurred from recontamination after cooking. In either

case, when food poisoning has occurred the contaminated, cooked carcasses had been held for many hours at temperatures which favoured the multiplication of the bacteria to a level whereby they caused the disease (Todd, Pivnick, Hendricks, Thomas and Riou, 1970).

For the routine detection of these organisms, a large sample is required and it has been the practice to remove the meat aseptically from one half of each chicken and homogenize it with twice its weight of sterile diluent. In general surveys where the carcasses cannot be sacrificed, swabs of a large area of the surface are taken.

Poultry rolls etc.

The use of cooked poultry by the small caterers was in the past hampered by the relative inconvenience of the whole bird in terms of cooking and serving. This disadvantage has been overcome in the poultry industry by the development of precooked poultry rolls, loaves and roasts. With rolls, pieces of meat are placed in a fibrous moisture proof casing and the rolls are cooked, cooled and stored. Poultry loaves are prepared in ham presses and, on cooling, packed in heat shrunk film. Roasts are formed by wrapping the deboned meat in the skin and tying it in such a way that its shape resembles that of the whole bird. These products are generally frozen.

The organisms most likely to survive cooking are the *Clostridium* spp and *Bacillus* spp and the thermoduric non-sporing bacteria such as faecal

FIG. 3. Sampling thawed poultry rolls along the central axis using a sterile cork borer.

BÜCHLI, K. (1965). Microbiological control on poultry processing plants with the "ten Cate Sausage". *Bull. Int. Inst. Refrig., Annex* (1), 227.

TEN CATE, L. (1965). A note on a simple and rapid method of bacteriological sampling by means of agar sausages. *J. appl. Bact.*, **28**, 221.

CLARK, D. S. (1965a). Method of estimating the bacterial population on surfaces. *Can. J. Microbiol.*, **11**, 407.

CLARK, D. S. (1965b). Improvement of spray gun method of estimating bacterial populations on surfaces. *Can. J. Microbiol.*, **11**, 1021.

DREWNIAK, E. E., HOWE, M. A., GORESLINE, H. E. & BAUCH, E. R. (1954). Studies on sanitizing methods for use in poultry processing. *U.S. Dept. of Agriculture. Circular* 930.

FOTER, M. J. & RAHN, O. (1936). Growth and germination of bacteria near their minimum temperature. *J. Bact.*, **32**, 485.

FROMM, D. (1959). An evaluation of techniques commonly used to quantitatively determine the bacterial population on chicken carcasses. *Poult. Sci.*, **38**, 887.

GALTON, M. M., MACKEL, D. C., LEWIS, A. L., HAIRE, W. C. & HARDY, A. V. (1955). Salmonellosis in poultry and poultry processing plants in Florida. *Am. J. vet. Res.*, **16**, 132.

GORESLINE, H. E. & HAUGH, R. R. (1959). Approximation of surface areas of cut-up chicken, and use in microbiological analysis. *Fd Technol., Champaign*, **13**, 241.

HAINES, R. B. (1931). The influence of temperature on the rate of growth of *Sporotrichum carnis* from $-10°C$ to $+30°C$. *J. exp. Biol.*, **8**, 379.

HAINES, R. B. (1934). The minimum temperatures of growth of some bacteria. *J. Hyg., Camb.*, **34**, 277.

HOBBS, B. C. (1971). Food poisoning from poultry. In *Poultry Disease and World Economy* (R. F. Gordon and B. M. Freeman, eds). Edinburgh: British Poultry Science Ltd.

KOTULA, A. W. (1966). Variability in microbiological samplings of chickens by the swab method. *Poult. Sci.*, **45**, 233.

LAHELLEC, C. & MEURIER, C. (1970). Variations dans l'importance quantitative des contaminations bactériennes des carcasses de volailles. *Bull. d'Information, Station Experimentale d'Aviculture de Ploufragan (Côtes-du-Nord)*, **10**, 118.

LEISTNER, L. & SZENTKUTI, L. (1970). Two methods for the bacterial examination of poultry carcasses. *Die Fleischwirtschaft*, **50**, 81.

MEAD, G. C. & IMPEY, C. S. (1970). The distribution of clostridia in poultry processing plants. *Br. Poult. Sci.*, **11**, 407.

MERCURI, A. J. & KOTULA, A. W. (1963). Relation of "Breast swab" to "Drip" bacterial counts in tray packed chicken fryers. *J. Food Sci.*, **29**, 854.

MICHENER, H. D. & ELLIOTT, R. P. (1964). Minimum growth temperatures for food poisoning, faecal indicator and psychrophilic microorganisms. *Adv. Fd Res.*, **13**, 349.

PATTERSON, J. T. (1971). Microbiological assessment of surfaces. *J. Fd Technol.*, **6**, 63.

SIMONSEN, B. (1971). Methods for determining the microbial counts of ready-to-cook poultry. *World's Poult. Sci. J.*, **27**, 368.

SURKIEWICZ, B. F., JOHNSTON, R. W., MORAN, A. B. & KRUMM, G. W. (1969). A bacteriological survey of chicken eviscerating plants. *Fd Technol., Champaign*, **23**, 1066.

TODD, E., PIVNICK, H., HENDRICKS, S., THOMAS, J. & RIOU, J. (1970). An evaluation of public health hazards of barbecued chickens. *Can. J. pub. Hlth*, **61,** 215.

WILLIAMS, M. L. B. (1967). A new method for evaluating surface contamination of raw meat. *J. appl. Bact.*, **30,** 498.

WOODBURN, M. & STADELMAN, W. J. (1968). Salmonellae contamination of production and processing facilities for broilers and ducklings. *Poult. Sci.*, **47,** 777.

ZIEGLER, F., SPENCER, J. V. & STADELMAN, W. J. (1954). A rapid method for determining spoilage in fresh poultry meat. *Poult. Sci.*, **33,** 1253.

Sampling and Estimation of Bacterial Populations in the Aquatic Environment

VERA G. COLLINS AND J. G. JONES

Freshwater Biological Association, The Ferry House, Ambleside, Westmorland, England

MARGARET S. HENDRIE AND J. M. SHEWAN

Torry Research Station, 135 Abbey Road, Aberdeen AB9 8DG, Scotland

AND

D. D. WYNN-WILLIAMS AND MURIEL E. RHODES

Department of Botany and Microbiology, University College of Wales, Aberystwyth, Wales

Introduction

A variety of apparatus has been devised and described for collecting samples of both the water column and bottom deposits for microbiological examination. The fact that new devices continue to be described or old ones modified suggests that most have drawbacks. The type of apparatus used will, to a certain extent, be determined by the sampling location and conditions and also by the information sought as a result of the sampling. Problems involved in sampling were discussed at the Interdisciplinary Conference at Princeton, January 1966 (Marine Biology, 1968).

Requirements of a good sampling device

In general, the collecting apparatus should meet the following requirements:

(a) It should be robust so as to withstand the rough handling normal on board ship and the high pressures to which it will be subjected at great depths.

(b) It should be capable of being sterilized, although it is not agreed that this is essential.

(c) It should be constructed of an inert material, i.e. not exert any bacteriostatic or bactericidal effect as can happen with some metals or certain types of rubber.

(d) It should be capable of collecting a sufficient volume of material for both microbiological and chemical analyses. Probably about 1 litre would be ample.

Examples of Sampling Apparatus

Water samplers

The Nansen water bottle

This sampler is of open-ended construction, being closed after it reaches the required depth, and is made of brass. It cannot be sterilized but can withstand high pressures. Kriss (1962, 1963) and Kriss, Lebedeva and Tsiban (1966) consider that it is satisfactory; as it passes quickly through the water column there is virtually no contamination and the water is in contact with the metal for such a short time that any bactericidal effect is barely noticeable. ZoBell (Marine Biology, 1968) and Sorokin (1964) refute these assertions and do not advocate the use of metal containers for sampling sea water for microbiological analysis. Sorokin (1960) has devised another type of deep water sampler which, it is claimed, avoids the criticisms levelled against the Nansen bottle. We have no experience with this apparatus.

The "Friedinger" water bottle and the "Ruttner" water sampler

The "Friedinger" water bottle (manufactured by Hans Büchi, Berne, Switzerland) is a stainless steel open-ended bottle which can be closed by means of a dropweight ("messenger") which slides down the supporting winch-wire or nylon cord. It is available in different sizes, the 1 litre bottle being the standard choice for sampling lakes.

The "Ruttner" water sampler is also an open-ended device (manufactured by Hydro-Bios Apparatebau GmbH, 23 Kiel-Holtenau, Germany). It is also available in various capacities and is constructed of either glass or Plexiglass.

The "Friedinger" bottle has a distinct advantage in its closing mechanism, in that minimum disturbance is achieved by the simple expedient of closing a top and bottom lid thereby enclosing the sample of water in a steel cylinder. The closing mechanism of the "Ruttner" bottle produces a turbulent flow-in due to suction resulting in considerable disturbance within any discrete zonation at the depth sampled. Another advantage that the "Friedinger" bottle has over the "Ruttner" sampler is

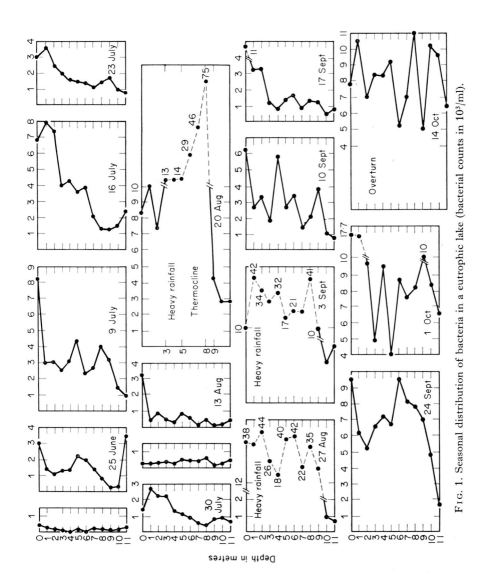

FIG. 1. Seasonal distribution of bacteria in a eutrophic lake (bacterial counts in 10³/ml).

that it allows an unimpeded flow of water through the sampler on lowering, whereas there is a delay of flow-through of water in the "Ruttner" bottle thereby increasing the hazard of cross contamination between sampling layers. Criticism levelled at the use of the "Friedinger" sampler (as with the Nansen bottle) suggests that contamination of the water samples occurs by the carry down of bacteria from water layers above to subsequent depths in a profile sampling series and therefore the apparatus is not suitable for bacteriological sampling. Results obtained from a series of 1 m depth pro-file samples using a "Friedinger" bottle of 1 litre capacity are illustrated in Fig. 1. A stratified freshwater lake (Collins, 1970) provides an ideal environ-mental site for testing the efficiency of this sampling apparatus. It is evident from these results that there is, in fact, very little "carrydown" or cross contamination between 1 m depth samples. There is negligible adherence of suspended particles and bacteria to the smooth internal walls of the stainless steel cylinder of the "Friedinger" sampler.

These findings are also supported by recent results obtained by one of us (J. G. Jones) in the course of carrying out routine sampling on a stratified lake. Samples taken between 2 and 8 m contained 6000 to 8000 bacteria ml^{-1}; approximately 90% of this population consisted of a morphologically distinct bacterium which produced white mucoid colonies on CPS medium (Collins and Willoughby, 1962). This organism was not detected at depths greater than 8 m where the population of viable bacteria was completely different. No carry through was noted even when sampling at depth inter-vals of 0·5 m below the 8 m level despite the presence of large numbers of the distinct bacterial colony type in the sample immediately above a depth of 8·5 m.

It should also be noted that the "Friedinger" bottle takes a sample of water of known dimensions and depth whereas evacuated or partially evacuated samplers draw in a "cone" of water whose depth and dimensions cannot be accurately determined.

The J–Z sampler

This sampler, first described by ZoBell (1941*b*) consists of a stout glass bottle (beer bottles of 500 ml capacity are suitable) fitted with a rubber bung through which passes a small bore glass tube to which is attached a pressure tube with a sealed narrow bore glass tube at the other end (Fig. 2). After autoclaving for 15 min at 121° the bung is tightened immediately, creating a partial vacuum as the bottle cools. The bottle is fitted to a metal frame which is then attached to a hydrographic cable and lowered to the required depth. When in position, a messenger is released down the cable, breaking the sealed tube by striking the lever and causing the rubber tube

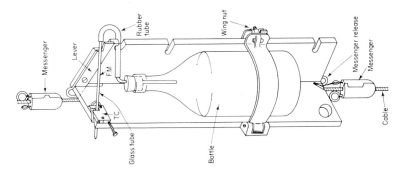

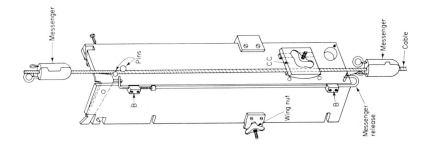

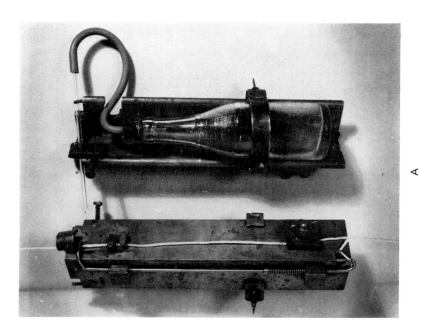

Fig. 2. The J–Z sampler (from ZoBell, 1941b).

to straighten and collect water well away from the cable and frame. As the apparatus is raised to the surface, air is expelled from the bottle due to decreased hydrostatic pressure, thus preventing contamination from the intervening water column. A series of samples at different depths can be taken by attaching a series of frames to one cable with messengers being released from the lower end of each frame as the messenger from above strikes the frame. A rubber bottle can also be used with this apparatus.

The J–Z sampler has been used successfully for sampling water in the North Sea and the fishing grounds of the Arctic and Faroe. Disadvantages of the apparatus are that it cannot be used at great depths and the volume of water collected does not usually exceed half the capacity of the bottle.

A modification of this apparatus for sampling anoxic organisms has recently been described (Schegg, 1970).

The Niskin sampler

This sampler (Niskin, 1962) was devised to enable large quantities of water to be sampled. It consists of a hinged frame of anodized aluminium or stainless steel which operates like a spring bellows (Fig. 3). A sterile plastic bag (made from 0·004 gauge stock tubular polythene film) of *c.* 3 litre capacity is fitted to the closed frame. When in position a messenger is released and strikes the knife which cuts the sealed tube leading into the bag. At the same time the catch holding the wings together is released, the torsion spring opens the wings drawing water into the bag. After filling, a clamping device closes off the tube so that the sample can be raised to the surface without contamination.

We have had a little experience with this apparatus, but have found difficulty in achieving a clean cut with the knife and an unimpeded flow of water into the bag. The designers claim that this difficulty has now been overcome. ZoBell (Marine Biology, 1968) lists four basic objections to the Niskin sampler:

(a) it is difficult to manipulate on board ship during rough weather;

(b) the plastic bags leak due to faulty seaming or pinholes or accidental puncture;

(c) its hydrodynamic behaviour makes it difficult to operate a series of samplers on one cast; and

(d) the water samples are contaminated from the cable etc. at the site of the cutting mechanism. This last objection, as has been indicated by Jannasch and Maddux (1967), is, in fact, true of most samplers, where contamination with the oil and other fluids from the suspending cable and metal frame may readily occur.

Jannasch and Maddux (1967) have devised a sampler in which the water is drawn from an area well away from any contamination carried by the cable or the sampler itself. Using tracer organisms they claim that their sampler collects water which is virtually free of contamination from the exterior of the equipment, whereas there is frequently contamination of the sample with the Niskin or the J–Z apparatus. However, this sampler is an adapted 50 m syringe and will presumably collect only a fairly small volume of water.

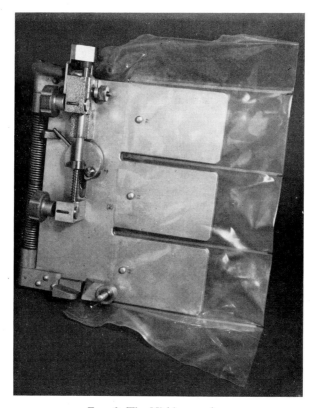

FIG. 3. The Niskin sampler.

The mouse-trap or rat-trap bottle

This simple, inexpensive apparatus (suggested by Professor de la Roy, Poitiers, pers. comm.) has been successfully used at Aberystwyth to sample surface waters in Cardigan Bay and in an Icelandic fjord (Report,

1970). For depths to 10 m, autoclaved, corked 500 ml medical flat bottles were screwed to a triggered mouse-trap (Fig. 4) so that when the trap was released, the cork jerked out of the bottle. When this was done under-water, the bottle filled with water and was automatically sealed by a second closure cork within the bottle, which floated up on the surface of the incoming water. When in use the bottles were clamped to a weighted support, lowered from the boat to the required depth and the mouse-trap was released by a messenger. For depths up to 15 m the more robust 500 ml blood plasma bottles were fitted to similarly triggered rat-traps.

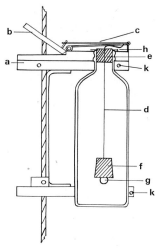

FIG. 4. The rat-trap bottle. a, Rat-trap; b, trigger platform; c, trigger arm; d, cord attached to upper greased bung, e, and then to lower greased bung, f, which has a glass weight, g, to ensure that it remains in the correct position to close the neck of the bottle. h, Anti-pressure pin to prevent bung, e, being forced into the bottle. The whole apparatus is attached to the cable with clamps, k.

For coastal work these bottles were satisfactory and were easy to handle in a small boat. They could be sterilized, were inexpensive and the sample was ready for examination without further transfer. Their main limitation was that the depth which could be sampled was restricted to 15 m or less by hydrostatic pressure.

The Mortimer sampler

A bacteriological sampler (Fig. 5) that provides sterile sampling *in situ* which eliminates the need to pre-evacuate the sterilized sampling bottle is that designed by Mortimer (1940).

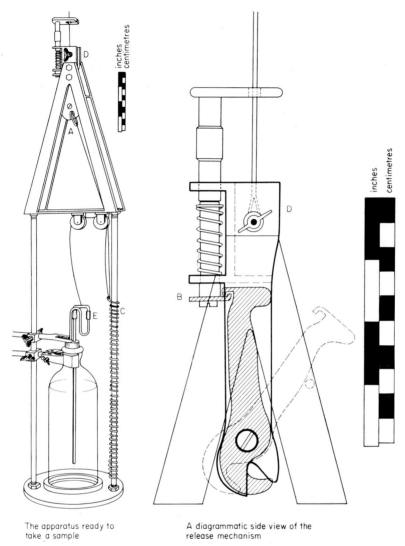

The apparatus ready to
take a sample

A diagrammatic side view of the
release mechanism

FIG. 5. The Mortimer apparatus. From *J. Hyg., Camb.* (Mortimer, 1940).

The true sample water bottle and the Bathyrophe

ZoBell (Marine Biology, 1968) has suggested that the "True Sample Water Bottle"—a modification of the J–Z sampler which can collect at least 1·5 l of water—might satisfy most demands. It consists of a heavy rubberized cylinder which collapses when evacuated and which can be autoclaved. It is difficult to see how this and similar rubberized containers, such as the

Bathyrophe (Brisou, 1965) and the piggy-back device for use in conjunction with the Nansen bottle (Sieburth, Frey and Conover, 1963), avoid the objections already raised and to which might be added a further one viz. the loss of elasticity or resilience at the high pressures and low temperatures at depths greater than 6000 m. However the Bathyrophe has been used extensively and apparently successfully by Brisou and his school in both shallow and deep-sea sampling.

The Williams membrane filter apparatus

Apart from the technical difficulties in collecting water samples already mentioned, other factors may well affect the final microbiological analysis. With most deep water samplers already mentioned the container has to be hauled on board ship where further manipulations are carried out. If taken from considerable depths, the decrease in pressure results in an exchange of gases in the container as an equilibrium is reached which is quite different from that at the sampling point. Consequently the microbial flora may then be in an ecosystem so different from that at which it was collected that many components may be unable to survive. The sampling apparatus is raised through surface waters where the temperature may be considerably higher than at the sampling point. This could result in extreme or obligate psychrophiles being killed. Similarly, obligate barophiles would not be expected to survive the decrease to atmospheric pressure when the sample is hauled on board ship. Finally, the surface water surrounding the ship may be heavily contaminated from the film of material adhering to the ship's hull or by sewage from the ship, although some research ships discharge polluting effluent only at certain times so as to minimize the risk of contamination during sampling.

In order to eliminate some of these problems and sample, as far as possible, *in situ*, Williams (1969) has developed a membrane filter apparatus which can be used at great depths. It is available from Hydro-Bios Apparatebau GmbH. After lowering to the required depth, a messenger breaks the seal allowing water to enter the cylinder through the membrane. A non-return valve prevents the water from flooding the membrane as the apparatus is raised. This apparatus eliminates the need for much of the immediate handling work on board ship. All that is necessary is the transfer of the membrane to a suitable growth medium and the measurement of the volume (at atmospheric pressure) of water which has been filtered through the membrane into the cylinder. It has the disadvantage that initially only one microbiological operation can be performed on any individual sample.

Monitoring apparatus

Oxygen tension

The main instrument used for the purpose of monitoring the environment at the sampling site is the Mackereth oxygen electrode (Mackereth, 1964). This instrument (obtainable from The Lakes Instrument Co. Ltd., Oakland, Windermere, Westmorland) simultaneously records temperature and dissolved oxygen. A typical series of determinations taken with this instrument on a small stratified lake is shown in Fig. 6. These results provide environmental monitoring data to Fig. 1.

Temperature

Oppenheimer (Marine Biology, 1968) reports the use of a reversing thermometer attachment to the Niskin sampler which reverses only after the bag has filled and also gives an additional check on the depth.

Field monitoring for acetylene reduction

Millipore field monitors were used in the technique fully described in the Handling of Samples section.

Sediment samplers

Sediment samplers are of two types viz. the grab sampler which scoops up sediment from the bottom and the core sampler which is driven into the sediment and takes a core of sediment with or without overlying water. Sediment samples from the littoral and intertidal zones can be collected directly into sterilized containers (jars or plastic bags) using a sterilized spatula or spoon.

The Shipek sampler

This grab sampler can collect unconsolidated sediment ranging from soft ooze to hard-packed coarse sand at any depth. The sampling bucket is the inner part of a pair of concentric half cylinders and is rotated at high torque by two helically wound external springs. On contact with the bottom, the trigger mechanism is automatically released by a self-contained weight allowing the inner half cylinder to scoop up sediment. The bucket is held in the closed position during the return to the surface. The Model 8 Shipek sediment sampler which has been used at Torry Research Station is avail-

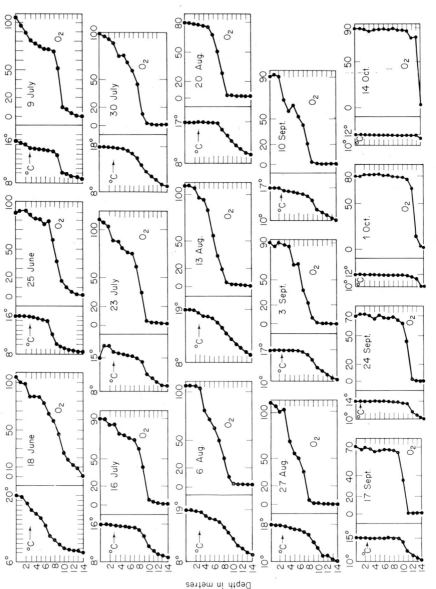

Fɪɢ. 6. Seasonal temperature (°C) and dissolved oxygen (% saturation) in a eutrophic lake.

able as Hydro Products Model 80 from Techmation, 85 Edgware Way, Edgware, Middlesex, HA8 8JP (Fig. 7). Difficulty was experienced in getting the sampler to go down straight, but this was rectified by attaching a plastic float at the upper end of the equipment (Davies, 1967).

This sampler is cumbersome and cannot be sterilized, but as the quantity of material collected is fairly large (0·04 m² to a depth of 0·1 m at the centre) it is possible to obtain some material which has had little contact with the equipment. It is not suitable for collecting samples in the top centimetre or so of the sediment.

The Jenkin surface-mud sampler

This core sampler (Fig. 8), designed for use in lakes, is fully described by Macan (1970) and is available from The Lakes Instrument Co. Ltd., Oakland, Windermere, Westmorland. The sampler collects a core of the surface mud together with the overlying water *in situ* in a relatively undisturbed state. The cores can be put to a variety of experimental uses which are outlined in the section on handling of samples.

The Craib sampler

This sampler, for collecting undisturbed cores of marine sediments, is fully described by Craib (1965). It is suitable for taking samples where it is desired to examine the microbial population at various depths within the sediments.

Media

The choice of medium to be used depends on the nature of the sample and information required from the sample. The medium may be a general one which will support the growth of the aerobic heterotrophic bacteria present in either freshwater or marine environments, or it may be selective in that it will enhance the growth of or demonstrate the presence of certain types of organisms, e.g. heterotrophic anaerobes, autotrophic types, fluorescent types, chitinoclastic bacteria, etc. The mineral requirements of marine bacteria are reviewed by MacLeod (1965) and media were discussed at the Interdisciplinary Conference (Marine Biology, 1968).

Media for freshwater bacteria

The standard plate count medium used in routine sampling of lakes is Collins casein-peptone-starch (CPS) medium (Collins and Willoughby,

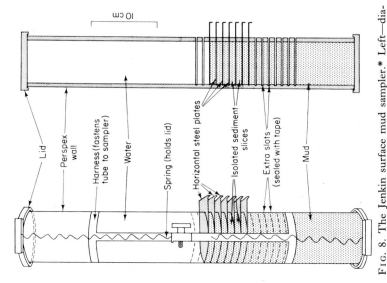

F I G. 8. The Jenkin surface mud sampler.* Left—diagrammatic representation of modified Jenkin sampler tube with horizontal steel plates in position isolating sediment slices. Right—vertical section of tube with plates in position (for greater clarity the harness is omitted and the thickness of the plates and slots is exaggerated).

* Illustration from Goulder (1971).

F I G. 7. The Shipek sampler.

1962). Full details of this medium are given in the Appendix. For pure culture maintenance, Collins' ENA medium (Collins, 1963) is used. For long-term maintenance of pure cultures, stab cultures in both CPS and ENA with the agar concentration reduced to 0.3% (w/v) dispensed in Bijou bottles is recommended.

The use of a modified CPS medium to estimate viable exo-enzyme producing bacteria has recently been described (Jones, 1971). Briefly, the basal CPS medium minus its starch and casein is used to enumerate protease, amylase and lipase producers by the addition of 0.5% gluten, 0.1% starch and 1.0% tributyrin respectively. Hot homogenization under aseptic conditions is required to produce a suitable emulsion in the lipase (tributyrin) medium. Clear zones developed around colonies producing protease or lipase; amylase zones were detected by developing with half strength Lugol's iodine. All media were adjusted to pH 7.0.

Media for marine bacteria

For the enumeration of marine bacteria a seawater based medium is required. Liston (1956) found that a seawater medium consistently gave higher counts of the bacteria from the skin and gills of freshly caught fish than did a meat extract medium of the type used in many laboratories. The bacterial types growing on a seawater medium are probably closely allied to the complete range of the flora of the environment, e.g. agar digesting bacteria were fairly often detected on seawater plates but not on the meat extract plates from the same specimen. When enumerating heterotrophic bacteria from waters of the North Sea, Anderson (1962) experimented with media of various compositions including a seawater version of nutrient agar, Medium 2216 (ZoBell, 1941a) and modifications of 2216. The media were prepared with seawater:distilled water in the ratio 3:1. He came to the conclusion that a modification of medium 2216 which had the peptone concentration reduced by a half and replaced by an equal concentration of yeast extract gave the highest count of bacteria. Chaina (1968) using various media, including Anderson's modification of 2216 found that a seawater Lemco agar (as used for fish work at Torry) gave the highest count. At Aberystwyth another modification of medium 2216—Johnson's (1968) Marine Agar (JMA)—was found satisfactory. A seawater yeast peptone medium has also been found satisfactory for growing luminous bacteria (Hendrie, Hodgkiss and Shewan, 1970) and for the maintenance of *Vibrio marinus*.

A medium for oligotrophic marine bacteria capable of growing at low nutrient levels has been devised at Aberystwyth. Details of this medium—

Aberystwyth Natural Sea Water Agar (AbSWA)—and the other media for marine heterotrophic bacteria are given in the Appendix.

Many media and techniques have been devised or adapted for the isolation, enumeration or selection of various groups of bacteria from the marine environment. A few examples are given here.

Clostridia

For the isolation of clostridia, Davies (1967, 1969) used Robertson's cooked meat medium prepared with either freshwater or seawater and incubated at 45, 30 or 20° for 3 days or 2·2° and 0° until growth occurred for enrichment. A cooked meat medium, but of different composition, was used for the isolation of psychrophilic clostridia by Liston, Holman and Matches (1969) but incubation was at 0 or 8°.

Alginate decomposing organisms

For the detection of alginic acid or alginate decomposing bacteria, methods have been devised by Kimura (1961) and Meland (1963).

Agarolytic bacteria

These can be detected by direct examination of plates for the general heterotrophic flora, or by staining the plates with iodine (Humm, 1946).

Chitin digesters

Chitin digesting bacteria are detected by the incorporation of chitin in a suitable medium (ZoBell and Rittenberg, 1938; Campbell and Williams, 1951; Chan and Liston, 1969).

Vibrio parahaemolyticus

Many media have been suggested for the detection of this organism such as BTB-Teepol, TCBS and AAC media (Baross and Liston, 1970, give full details), VPH-medium (Asakawa, Akabane and Noguchi, 1969 pers. comm. by H. Iida) or a high pH medium containing penicillin (Twedt and Novelli, 1971).

Luminous bacteria

Examination of count plates in the dark room for the detection and enumeration of luminous bacteria. Adequate time must be spent in the dark before examining plates to allow for accommodation of the eyes—10 min is sufficient.

Fluorescent pseudomonads

A seawater modification of Medium B (King, Ward and Raney, 1954) has been found useful at Aberystwyth for detection of fluorescent pseudo-monads in the marine environment. Details of this medium are given in the Appendix.

Thiobacilli

The isolation and enumeration of marine thiobacilli has been described by Tilton, Cobet and Jones (1967).

Actinomycetes

Despite earlier assertions that actinomycetes are fairly uncommon in the marine environment, Weyland (1969) found considerable numbers in sediments of the North Sea and Atlantic Ocean using seawater based media.

Yeasts

Various media have been used to isolate yeasts from the marine environment (Shinano, 1962; van Uden and Castelo-Branco, 1963; Fell and van Uden, 1963; Ross, 1963; Ross and Morris, 1964).

Autotrophs and nitrogen-fixing bacteria in the marine environment have been studied particularly at Aberystwyth in addition to the general hetero-trophic aerobic flora. For the detection of nitrogen-fixing bacteria, experiments were made with a seawater version of the "nitrogen free" medium of Brown, Burlingham and Jackson (1962) in which the addition of a small quantity of yeast extract to the medium was often necessary to obtain growth which could be subsequently subcultured. Attempts were made to test the final medium—Brown's Synthetic Sea Water Yeast Agar (BSSWYA)—with a known nitrogen-fixing bacterium, but it was not possible to obtain one of marine origin and terrestrial strains did not grow satisfactorily in the higher salt concentration of the medium.

A Marine Athiorhodaceae Medium (MAM) based on that of Haskins and Kihara (1967) and a Marine Thiorhodaceae Medium (MTM) based on Pfennig's (1965) were used for the enrichment of anaerobic photo-trophic bacteria from sediments and sea water. Full details of these media are given in the Appendix.

Handling the Samples

Samples should be "processed" as soon as possible after collection. ZoBell (1946) states that the numbers of bacteria in water samples change fairly rapidly after collection, an initial decrease in numbers being followed by a rapid increase. The change in total numbers is accompanied by a change in the relative proportions of the types present. If there is any unavoidable delay, the samples should, where possible, be refrigerated until they can be handled. Samples should be thoroughly shaken (e.g. by hand for 5 or 10 min) to ensure even distribution of the organisms throughout the sample before proceeding to dilute, if necessary, and plate.

When examining samples for obligately psychrophilic organisms, in addition to keeping the samples at a low temperature, the work should, if possible, be carried out in a chilled laboratory, or if such facilities are not available, all media, dilution blanks and equipment should be well chilled in a refrigerator before use and not allowed to stand at room temperature for any length of time.

Counting procedures

Counts of viable bacteria are carried out in one of three ways, as spread plates, as pour plates (or in roll tubes) or on membrane filters. Where the sample requires dilution, the diluent used will depend on the nature of the sample. Sterile lake water is used to dilute lakewater samples and sterile filtered seawater or artificial seawater to dilute marine samples. Sterile $3 \cdot 0 \%$ NaCl solution could also be used as a diluent for marine samples.

Spread plate procedure

A known volume (e.g. $0 \cdot 1$ or $0 \cdot 2$ ml) of sample or diluted sample is spread on the surface of an agar plate in a standard manner and incubated for a suitable period (e.g. 24 days/$10°$ has been found suitable for freshwater samples; 5 days/$20°$ or 25 days/$0°$ has been found necessary to give maximum colony counts on specimens from marine sources (J. Liston, pers. comm.)). When dealing with a large number of freshwater samples, the following variation of the technique was found to be time saving involving the spreading of smaller quantities of the water sample and dispensing with the preparation of dilution blanks. A known, but not accurately measured volume of sterile lake water (e.g. 1 drop delivered aseptically from a Pasteur pipette) is placed in the centre of an agar plate. Into this $0 \cdot 01$ ml of lakewater sample is pipetted and the mixture is spread in the normal manner. In this way it is possible to spread a small volume over the surface of an agar plate.

Anderson (1962) using volumes of 1·0 ml seawater sample spread on the surface of well dried plates (at 50° for 2–3 h) found that this volume could be absorbed within 30 min, but we generally prefer to use a smaller volume.

Pour plate procedure

A known volume (1·0 ml) of sample or a suitable dilution of the sample is pipetted into a sterile Petri dish. Molten agar medium, held at 46°, is added and the whole is mixed in a standard manner and incubated for a suitable period—10 days at 20° has been found suitable for lakewater samples.

An objection to this method is that it exposes all the organisms present in the sample to a temperature of 46°. Some psychrophilic types are not able to survive this temperature exposure. Bearing in mind the findings of Clark (1967, 1971) using pure cultures, spread plates are preferred to pour plates.

Viable counts on membrane filters

A suitable volume of sample (this can vary considerably with the body of water under examination) is drawn through a membrane filter and the membrane is placed on a suitable agar plate.

For lakewater samples a 0·22 μm membrane (Millipore UK Ltd) is used and is placed on CPS solid medium. After incubation (e.g. 2 days/10° or 36 h/15°) the membranes are removed, dried and stained with 1% aqueous erythrosin for 30 min. After further drying, the membrane is washed twice in distilled water, re-dried, cleared with immersion oil or cedarwood oil and examined microscopically. The colonies are stained pink and the excess stain has been removed from the membrane by the washing procedure. All staining and washing procedures are done through the membrane, allowing the solutions to soak through from below, and using only membrane filtered stains and water. Full details of this method and examples of its use are given in Jones (1972).

When dealing with seawater samples, Chaina (1968) used 0·45 μm membranes, 2·0 ml of samples were filtered and the membranes were incubated for 5 days at 20°. The colonies appearing on the membrane were counted directly. Three counts were made on each sample on each of the media used. By using the direct counting of colonies without staining, colonies can be selected for qualitative examination of the flora.

Sediment samples

Samples collected from lakes by means of the Jenkin surface-mud sampler can be put to a variety of experimental uses. For studies on the depth distribution of mud bacteria the overlying water of the sample is siphoned off, the mud core is pushed up the tube with a piston and slices are cut off at 1 cm intervals and placed in sterile Petri dishes for bacteriological examination. The mud core samples as collected make ideal Winogradsky cylinders for use as enrichment cultures for many selective groups of bacteria (Collins, 1969).

Studies on redox potential measurements can be readily carried out on the mud samples, either by the insertion of electrodes from the top of the tube, or by drilling holes of small diameter at 1 cm intervals through the sides of the Perspex tube and inserting microelectrodes made out of syringe needles into the drilled holes. This method can be used for discrete sampling of

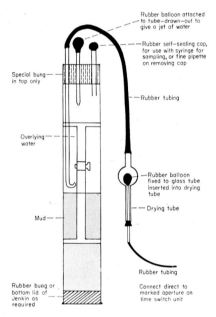

F I G. 9. Perfuser method for Jenkin surface mud cores. The Jenkin tube must be absolutely full of water. The two small tubes on the top bung must be filled with water. Attach the tubing from either a vacuum pump, or water filter pump to the appropriately marked aperture on Time Switch Unit. This switch unit is set to "make and break" contact once every minute approximately. The two rubber balloons will depress during this cut-out time, for a period of 5 sec. When the switch operates the balloons will eject water into the system, then refill as the switch makes contact again, hence stirring the core water, without introducing air to do so.

the mud by the insertion of sterile syringe needles at 1 cm intervals and by the application of suction, microsamples of mud from the various layers in the mud core can be withdrawn for bacteriological procedures. Also holes of a suitable diameter allow for the removal of samples at 1 cm depth intervals, by means of a cork borer, without the application of suction.

An alternative method of discrete sampling of the mud cores is to isolate 1 cm slices of the mud by means of cutting slots in the Perspex tube of the sampler at 1 cm intervals and by inserting steel plates into the slots thereby isolating discrete slices. Each isolated slice of mud is then scraped out with a long-handled scraper into suitable containers. For the application of this method to studies of microorganisms other than bacteria see Goulder (1971).

One of the main uses of mud samples obtained by the Jenkin surface-mud sampler is in respiration studies. Respirometric determinations can be performed *in situ* in the tubes of mud and by the use of a perfuser technique (Fig. 9) it being possible to keep the overlying water circulating without the introduction of air into the system. Alternatively samples of mud can be removed from the tubes by any of the methods previously described for direct determinations in a respirometer.

For the isolation of heterotrophic anaerobes from marine sediments, a portion of sample (*c.* 2 g) was added to Robertson's cooked meat broth and incubated at various temperatures and then plated out on various media recommended for clostridia (Davies, 1969) and incubated anaerobically at the appropriate temperature. Liston *et al.* (1969) used similar methods.

Enrichment and isolation of anaerobic photoautotrophic bacteria from marine mud

A modification of Winogradsky's (1888) mud column technique is suitable for crude enrichment cultures of photosynthetic bacteria. Approximately 1 litre samples of mud are blended with 4·0 g $CaSO_4$, 4·0 g $CaCO_3$ and 5·0 g shredded filter-paper pulp, and the mixture filled into 30–40 cm lengths of 32 mm diam "lay flat" plastic tubing (Transatlantic Plastics Ltd., Garden Estate, Ventor, Isle of Wight, UK). Care is taken to minimize air bubbles. These mud "sausages" are firmly tied and incubated at 20–24° at *c.* 30·0 cm distance from a 60 W tungsten-filament bulb. After about 14 days any coloured patches of presumptive photosynthetic bacteria may be removed by puncturing the sausages with sterile hypodermic needles, and thus transferring growth to suitable liquid marine Thiorhodaceae and Athiorhodaceae media. Liquid cultures are purified by inoculating suitable molten agar to make anaerobic shake cultures, and repeated selection of single colonies thus entrapped. Liquid or agar cultures are incubated anaerobically in full, sealed containers in a water-bath at 20–24°, illuminated from below by 60 W tungsten-filament bulbs at a distance of 20·0 cm to give a light intensity of *c.* 300 lux.

Acetylene reduction procedure

The ability of nitrogen-fixing bacteria to reduce acetylene by attacking its triple bond, was found by Dilworth (1966) and Schöllhorn and Burris (1966), to be correlated with the ability to rupture the triple-bonded nitrogen molecule, viz. to "fix" molecular nitrogen. This correlation was promptly exploited by Stewart, Fitzgerald and Burris (1967) in the field; they found acetylene-reducing blue-green algae in several Wisconsin lakes, but no presumptive nitrogen-fixing bacteria. Hardy, Holsten, Jackson and Burns (1968) developed an incubation chamber technique for field use, and this was further modified here for the detection of low densities of aerobic acetylene-reducing bacteria. Another group of potential nitrogen-fixing bacteria is the anaerobic photosynthetic bacteria, which have been reported from marine waters and sediments (Stewart, 1968, 1969; Loponitsina, Novozhilova and Kondrat'eva, 1969). Appropriate techniques were therefore developed also for the enrichment and pure culture study of marine Thiorhodaceae and Athiorhodaceae.

For aerobic heterotrophs isolated on nitrogen-poor BSSWYA. Pure cultures of bacteria which maintained good growth after several subcultures on Brown's nitrogen-poor synthetic seawater yeast extract agar, were grown aerobically on 2·0 ml slopes of this agar in Bijou bottles for 3–5 days at 22°. Fresh BSSWY broth (2·0 ml) was then added to provide a suitable additional carbon and energy source, and the cotton-wool plug was replaced by a sterile (121° for 20 min) "Subaseal vaccine cap" (Messrs Gallenkamp, PO Box 19, Victoria House, Croft Street, Widnes, Lancs., UK). Using hypodermic needles attached eventually to a gas cylinder, the bottles were then thoroughly flushed with a gas mixture of 95·0% argon:5·0% CO_2. Acetylene, generated from BDH calcium carbide with water, was then injected to give a 10·0% atmosphere over the culture, and the bottles were then agitated on an orbital shaker for 6 h at 24°.

After incubation, 0·5 ml of the gas phase over the culture was withdrawn using a hypodermic syringe, and injected into a Pye 104 gas chromatograph fitted with a flame ionization detector head at the end of a 3 m column packed with 85–100 mesh activated alumina (Jones Chromatography Co. Ltd., Newport, Glam., UK). Nitrogen was employed as the carrier gas at a flow rate of 45·0 ml per min and the oven was at 145°. Methane, ethylene and acetylene were detected on a Servoscribe chart recorder, and the screening time for each sample was about 3 min.

For anaerobic photosynthetic bacteria. To test for ability to reduce acetylene the purified cultures of photoautotrophic bacteria were grown anaerobically in sealed Bijou bottles containing 6·0 ml of MAM or MTM (medium details in Appendix). One ml of young cell suspension (age varied with isolate) was then transferred to a further "Subaseal"-capped Bijou bottle containing 2·0 ml of fresh nitrogen-free medium. The gas phase was replaced with 95% argon:5% CO_2, and acetylene finally added to give a 10·0% concentration. After incubation for 6·0 h in the illuminated water-bath at 20–24°, the gas phase was screened for the presence of ethylene as already described for heterotrophic isolates. Control cultures minus acetylene were essential for the detection of extraneously-produced hydro-carbons.

The direct use of field monitors for acetylene reduction studies

Preliminary results using spread plates of nitrogen-poor BSSWYA indicated very low numbers of potential nitrogen-fixing marine bacteria, so they were concentrated from the seawater by membrane filtration. Well-mixed sea-water (50·0 ml) was drawn into a plastic syringe via a size 0 serum needle, and injected into the top of a gas-tight Millipore Field Monitor containing a filter membrane of A.P.D. 0·45 μm, and fitted with a No. 2 "Chekaleke" soft rubber bung (obtainable from the International Bottle Co. Ltd., 140 Park Lane, London, W.1.). The bottom half of the monitor was fitted with a No. 4 "Chekaleke" bung, which could be penetrated by a No. 2 serum needle (see Fig. 10). Thus, either gases or liquids in either the upper or lower halves of the monitor could be manipulated by means of hypodermic syringes. The rubber bungs were gas-tight, and PVC adhesive tape was used to effect a gas-tight seal of the two monitor halves.

When preliminary growth and counting of the organisms retained on the membrane was required, 0·8 ml of BSSWY medium was injected into

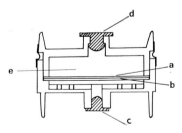

FIG. 10. Field monitor for acetylene reduction. Millipore Field Monitor with 37 mm membrane, pore size 0·45 μm (a) over cellulose pad (b). The two halves are sealed together with PVC tape and the openings are fitted with "Chekaleke" rubber bungs, (c) No. 4 through which medium is inserted and (d) No. 2 through which sample is inserted. (e) Gas space.

the cellulose pad immediately underneath the membrane, and after incubation in an aerobic gas phase for 5 days at 22°, colonies were counted using shallow-angle lateral illumination and a binocular microscope.

The ability to reduce acetylene by either anaerobically- or aerobically-incubated crude cultures, or non-incubated seawater residues, was tested by injecting 2·0 ml of BSSWY medium (i.e. a suitable fresh carbon and energy source for acetylene reduction) into the upper half of the monitor. After agitation to suspend the organisms, the gas phase was replaced with 95·0% argon : 5·0% CO_2; finally 10·0% of this gas volume was replaced by acetylene. Monitors were incubated on an orbital shaker at 24° for various periods of time prior to testing for the presence of ethylene in the gas phase as already described above for pure cultures.

The results thus provided data enabling comparisons between the size of an aerobic or anaerobic viable colony count and its magnitude of acetylene-reducing activity. It was essential to include suitable control monitors to detect spontaneous ethylene production from various sources.

The field monitors could also be used both to concentrate and enrich anaerobic photoautotrophic bacteria, and to test for their ability to reduce acetylene, because they could conveniently be supplied with suitable liquid media and gaseous environments. Such anaerobic monitors were floated in the illuminated water-bath at 20–24° to effect the necessary conditions for energy generation to fix nitrogen or reduce acetylene.

The advantages of the above procedure for screening potential nitrogen-fixing marine bacteria are that it provides a means of concentrating and enriching such aerobic or anaerobic aquatic bacteria in a non-dehydrating environment. Colony counts of organisms able to grow on a nitrogen-free medium may be obtained, and their acetylene-reducing ability as a natural association is detectable owing to their enclosure in a gas-tight chamber which may easily be injected with any desired nutrient or gas mixture, for either growth and/or acetylene reduction. Recovery of viable acetylene-reducing organisms is also possible. It was found that the monitors could be reused after washing, renewing membranes, and overnight sterilization in a 10·0% ethylene oxide : 90·0% CO_2 gas mixture.

Appendix

Media formulation and preparation

Casein-peptone-starch (CPS) medium

Soluble casein (BDH)	0·5 g
Bacto-peptone (Difco)	0·5 g
Soluble starch	0·5 g
K_2HPO_4	0·2 g
$MgSO_4.7H_2O$	0·05 g
$FeCl_3$	trace (4 drops of a 0·01 % solution)
Agar (Difco)	15·0 g
Glycerol	1·0 ml
Distilled water	1000 ml

The casein, peptone, starch and agar are added to water and dissolved by steaming. The remaining ingredients are added and the complete medium is mixed. Finally the molten medium is filtered on a Buchner funnel containing absorbent cotton wool, previously rinsed in distilled water and squeezed to remove excess water, the entire filtration unit having been brought to the same temperature as the medium in the steamer. The medium is sterilized at 121° for 20 min. The pH should be 6·9–7·0 without adjustment.

Basal medium for exo-enzyme studies

Bacto-peptone	0·5 g
K_2HPO_4	0·2 g
$MgSO_4.7H_2O$	0·05 g
$FeCl_3$	trace (see CPS above)
Agar	15·0 g
Glycerol	1·0 ml
Distilled water	1000 ml

To this basal medium the following compounds are added:

gluten	(5·0 g)	for protease
or starch	(1·0 g)	for amylase
or tributyrin	(10·0 g)	for lipase

Seawater

Seawater for media preparation should be collected at a point reasonably free from freshwater and terrestrial pollution. In order to obtain reproducible results it should be stored in the dark in a glass container for a number of weeks and then filtered to remove particulate matter (ZoBell,

1946). Stocks of seawater are stored for a minimum of 3 weeks and poly-thene containers are used. The water is filtered through paper (Green's "Hyduro" No. 904½ or Whatman No. 1). Experience over the years has shown that a salinity level of 25‰ is adequate for most marine bacteria and, in order to minimize precipitation during sterilization, the seawater is diluted with distilled or deionized water (3 parts seawater:1 part distilled or deionized water) for the majority of media prepared.

Artificial seawater may also be used for media preparation. This may be prepared from a fairly simple formula, such as that used at Aberystwyth—NaCl, 27·0 g; $MgCl_2$, 2·5 g; KCl, 0·75 g/l of deionized water, or a more complex formulation based on the chemical composition of seawater (Lyman and Fleming, 1940). MacLeod, Onofrey and Norris (1954) give a suitable formula, based on Lyman and Fleming. Commercially prepared marine mixes are also suitable for media preparation.

Medium 2216 (ZoBell, 1941*a*)

Bacto-peptone	5·0 g
$FePO_4$	0·1 g
Bacto agar	15·0 g
Aged seawater	1000　ml
Final pH 7·5–7·6	

This medium was later modified by the addition of 1 g yeast extract and is known as medium 2216E. A dehydrated version (Bacto Marine Agar 2216E) for preparation with distilled water is available from Difco Laboratories (PO Box 148, Central Avenue, West Molesey, Surrey).

Anderson's medium (1962)—a modification of 2216

Peptone	2·5 g
Yeast extract	2·5 g
$FePO_4$	0·1 g
Agar	15·0 g
Aged seawater	750　ml
Distilled water	250　ml
Final pH 7·4–7·6	

Johnson's marine agar (1968)—a modification of 2216

Bacto-peptone	5·0 g
$FeSO_4.7H_2O$	0·2 g
$Na_2S_2O_3$	0·3 g
Yeast extract (Difco)	1·0 g
Bacto agar	15·0 g
Seawater	750　ml
Distilled water	250　ml

Seawater Lemco agar

Lab Lemco	10·0 g
Peptone	10·0 g
Agar	15·0 g
Aged seawater	750 ml
Distilled water	250 ml
Final pH *c.* 7·4	

Seawater yeast peptone agar (Hendrie *et al.*, 1970)

Yeast extract powder	3·0 g
Peptone	5·0 g
Agar	15·0 g
Aged seawater	750 ml
Distilled water	250 ml
Final pH *c.* 7·4	

Seawater King's agar (SWKA)

The medium B of King *et al.* (1954) is made with either natural aged seawater, or in synthetic seawater (see p. 102 for details), but with the phosphate concentration reduced to 0·2 g/l. Thus it contains:

Proteose peptone	20·0 g
Glycerol	10·0 g
K_2HPO_4 (anhyd)	0·2 g
$MgSO_4.7H_2O$	1·5 g
Oxoid agar No. 3	15·0 g
Aged seawater	750 ml
Deionized water	250 ml
Adjust to pH 7·2 and sterilize at 115° for 15 min.	

Brown's synthetic seawater yeast agar (BSSWYA) for the enrichment of potential nitrogen-fixing bacteria

The findings of Brown *et al.* (1962) were borne in mind when modifying their *Azotobacter* medium for marine bacteria, and 3 separately-sterilized solutions are still essential. They are:

Solution (A)		
	$MgSO_4.7H_2O$	0·2 g
	$FeSO_4.7H_2O$	0·04 g
	$Na_2MoO_4.2H_2O$	0·005 g
	$CaCl_2$ (anhyd)	0·15 g
	0·5 % Difco yeast extract	5·0 ml
	Oxoid Agar No. 3	15·0 g
	Synthetic seawater	995·0 ml

This medium is filled in 80·0 ml volumes in 4 oz. medical flat bottles and autoclaved at 115° for 15 min, when the pH value should be 8·1.

Solution B. Sterile (115° for 15 min) 0·2% K_2HPO_4 in deionized water.

Solution C. Sterile (115° for 15 min) 5·0% glucose in deionized water.

To 80·0 ml of molten, cooled (45°) solution (A) is added 10·0 ml each of solutions (B) and (C), and after mixing the plates are poured and dried.

Aberystwyth natural seawater agar (AbSWA)

Four separate solutions are required:

Solution A.	Aged seawater	675 ml
	Deionized water	225 ml
	Oxoid No. 1 agar	10·0 g
	(or 15·0 g Oxoid agar No. 3)	

After steaming to dissolve this is bottled in 190·0 ml volumes,

Solution B.	12·8% aq solution of	
	$FeSO_4{\cdot}7H_2O$	0·1 ml
	Na_2EDTA	0·04 g
	Deionized water	15·0 ml
Solution C.	Trace element solution*	0·1 ml
	Deionized water	14·9 ml
Solution D.	Na_2HPO_4	0·03 g
	Deionized water	70·0 ml

All 4 solutions are sterilized by autoclaving at 115° for 15 min. When required, to each bottle (190 ml) of molten medium A (steamed, *not* autoclaved, and cooled to 45°) is added, in order, 3·0 ml of solution (B), 3·0 ml of (C) and 4·0 ml of solution (D). A small precipitate forms, but if the medium is well-mixed throughout it is not sufficient to obscure even small colonies.

Marine Athiorhodaceae medium (MAM)

This medium is based on that of Haskins and Kihara (1967), and must be prepared in 5 parts, as follows:

Solution A.	$CaCl_2$ (anhyd)	380 mg
	Ferric citrate	50·0 mg
	$NaHB_4O_7$	20·4 mg
	$MnSO_4.4H_2O$	20·3 mg
	$ZnSO_4.7H_2O$	2·2 mg
	Deionized water	1000 ml
	Autoclaved at 115° for 15 min.	

---

* This contains: $MnSO_4.4H_2O$, 14·75 g; $FeSO_4.7H_2O$, 64·0 g; $CuSO_4.5H_2O$, 0·23 g; $CoCl_2.6H_2O$, 0·64 g; $Al_2(SO_4)_3.16H_2O$, 0·23 g; $LiCl.H_2O$, 0·052 g; $ZnSO_4.7H_2O$, 49·8 g, and $Na_2MoO_4.2H_2O$, 2·35 g in deionized water, 500·0 ml.

Solution B.　　CoSO$_4$.7H$_2$O　　　　　　　　9·6 mg
　　　　　　　CuSO$_4$.5H$_2$O　　　　　　　　7·8 mg
　　　　　　　Na$_2$MoO$_4$.2H$_2$O　　　　　　5·3 mg
　　　　　　　Deionized water　　　　　1000 ml
　　　　　　　　Autoclaved at 115° for 15 min.

Phosphate solution.
　　　　　　　2·0% K$_2$HPO$_4$ in deionized water.
　　　　　　　　Autoclaved at 115° for 15 min.

Biotin solution. The stock solution contains 5·0 mg per 100·0 ml deionized water; sterilized by membrane-filtration.

Basal medium.　Monosodium glutamate　　　4·0 g
　　　　　　　DL-Malic acid　　　　　　　3·5 g
　　　　　　　Na citrate　　　　　　　　0·8 g
　　　　　　　Difco yeast extract　　　　0·5 g
　　　　　　　MgSO$_4$.7H$_2$O　　　　　　0·2 g
　　　　　　　NaCl　　　　　　　　　　27·0 g
　　　　　　　MgCl$_2$　　　　　　　　　2·28 g
　　　　　　　KCl　　　　　　　　　　0·75 g
　　　　　　　Deionized water　　　　　934·0 ml

Adjust pH to 7·8–8·0 with KOH, and after sterilization at 115° for 15 min this basal medium should have a pH value of 7·6.

　The complete medium is made by the aseptic addition of 50·0 ml of solution (A), 5·0 ml of solution (B), 10·0 ml of the phosphate solution and 1·0 ml of the biotin solution to 934·0 ml of the basal medium. If desired, 15·0 g/l of Oxoid No. 3 agar may be incorporated in the basal medium.

Marine Thiorhodaceae medium (MTM)

This medium was based on that used by Pfennig (1965) for the enrichment of Thiorhodaceae and Chlorobacteriaceae from seawater. Six separate solutions are necessary:

Solution A. 0·08% anhydrous CaCl$_2$ in distilled water.

This is filled out at 11·0 ml aliquots in 18·0 ml screw-capped Bijou bottles, or as 18·0 ml aliquots in 30·0 ml McCartney bottles. Some residual solution for "topping up" is required. All sterilized at 115° for 10 min.

Solution B. Trace elements solution (based on Pfennig and Lippert, 1966)

Na$_2$EDTA	5·0 g	NiCl$_2$.6H$_2$O	0·02 g
H$_3$BO$_3$	0·03 g	Na$_2$MoO$_4$.2H$_2$O	0·03 g
ZnSO$_4$.7H$_2$O	0·1 g	KBr	0·048 g
MnCl$_2$.4H$_2$O	0·3 g	SrCl$_2$	0·012 g
CoCl$_2$.6H$_2$O	0·2 g	Na$_2$SO$_4$	1·809 g
CuCl$_2$.2H$_2$O	0·01 g	Glass-distilled water	1000 ml

Solution C. Vitamin B_{12} Solution. 2·0 mg Vitamin B_{12} (Cyanocobalamin, Merck) is dissolved in 100·0 ml glass-distilled water.

Solution D. To 15·0 ml glass distilled water is added 16·0 ml of the trace elements solution (B) and 1·0 ml of the vitamin solution (C), and also:

	$MgCl_2.6H_2O$	5·27 g
	KCl	0·55 g
	$FeSO_4.7H_2O$	0·028 g
	NH_4Cl	0·33 g
Solution E.	$NaHCO_3$	1·5 g
	NaCl	27·0 g
	KH_2PO_4	0·2 g
	Distilled water	300·0 ml

Carbon dioxide is bubbled into this solution for about 30 min until the pH value has fallen to *c.* 6·2. As soon as this solution is thus CO_2-saturated, solution (D) is added, and the mixture immediately sterilized by membrane filtration prior to adding it to the sterile bottled solution (A).

Solution F. 0·75 g $Na_2S.9H_2O$ is dissolved in 67·0 ml distilled water, autoclaved at 115° for 15 min, and then 1·5 ml sterile neutralizing 2 M-H_2SO_4 added dropwise, with constant agitation (magnetic stirring if possible). Note that the concentration of Na_2S may be varied to select for different photosynthetic bacteria. More Na_2S solution may be required as it is metabolized, but it must always be partially-neutralized first to a pH value of 8·0–8·5.

The final growth medium is aseptically combined in screw-capped containers as follows:

	McCartney bottle (30·0 ml)
Solution A	18·0 ml
Final combined solutions D and E	10·0 ml
Solution F	1·5 ml

Then solution A is used to "top up" the bottles, leaving a pea-sized air bubble to allow for pressure fluctuation.

References

ANDERSON, J. I. W. (1962). Heterotrophic bacteria of North Sea water. *Ph.D. Thesis*. Glasgow: University of Glasgow.

ASAKAWA, Y., AKABANE, S. & NOGUCHI, M. (1969). *Modern Media*, **15**, 234 (pers. comm. by H. Iida, 1970).

BAROSS, J. & LISTON, J. (1970). Occurrence of *Vibrio parahaemolyticus* and related hemolytic vibrios in marine environments in Washington State. *Appl. Microbiol.*, **20**, 179.

BRISOU, J. (1965). Le Bathyrophe. Appareil destiné a prélever aseptiquement les eaux aux grandes profondeurs. *Cah. océanogr.*, **17**, 53.

BROWN, M. E., BURLINGHAM, S. K. & JACKSON, R. M. (1962). Studies on Azotobacter species in soil. 1. Comparison of media and techniques for counting Azotobacter in soil. *Plant and Soil*, **17**, 309.

CAMPBELL, L. L. & WILLIAMS, O. B. (1951). A study of chitin-decomposing microorganisms of marine origin. *J. gen. Microbiol.*, **5**, 894.

CHAINA, P. N. (1968). A study of the bacterial flora, bacteriophages and a bacteriocin-like agent isolated from sea water and sea weeds collected at selected stations in the North Sea. *Ph.D. Thesis*. Aberdeen: University of Aberdeen.

CHAN, J. G. & LISTON, J. (1969). Chitinoclastic bacteria in marine sediments and fauna in Puget Sound. *Bact. Proc.*, p. 36.

CLARK, D. S. (1967). Comparison of pour and surface plate methods for determination of bacterial counts. *Can. J. Microbiol.*, **13**, 1409.

CLARK, D. S. (1971). Studies on the surface plate method of counting bacteria. *Can. J. Microbiol.*, **17**, 943.

COLLINS, V. G. (1963). The distribution and ecology of bacteria in freshwater. *Wat. Treat. Exam.*, **12**, 40.

COLLINS, V. G. (1969). In *Methods in Microbiology*, Vol. 3B (J. R. Norris and D. W. Ribbons, eds). London and New York: Academic Press.

COLLINS, V. G. (1970). Recent studies of bacterial pathogens of freshwater fish. *Wat. Treat. Exam.*, **19**, 3.

COLLINS, V. G. & WILLOUGHBY, L. G. (1962). The distribution of bacteria and fungal spores in Blelham Tarn with particular reference to an experimental overturn. *Arch. Mikrobiol.*, **43**, 294.

CRAIB, J. S. (1965). A sampler for taking short undisturbed marine cores. *J. Cons. perm. int. Explor. Mer.*, **30**, 34.

DAVIES, J. A. (1967). Clostridia from North Sea sediments. *Ph.D. Thesis*. Aberdeen: University of Aberdeen.

DAVIES, J. A. (1969). Isolation and identification of clostridia from North Sea sediments. *J. appl. Bact.*, **32**, 164.

DILWORTH, M. J. (1966). Acetylene-reduction by nitrogen-fixing preparations from *Clostridium pasteurianum*. *Biochim. biophys. Acta*, **127**, 285.

FELL, J. W. & VAN UDEN, N. (1963). Yeasts in marine environments. In *Symposium on Marine Microbiology* (C. H. Oppenheimer, ed.). Springfield, Illinois: Charles C. Thomas.

GOULDER, R. (1971). Vertical distribution of some ciliated protozoa in two freshwater sediments. *Oikos.*, **22**, 199.

HARDY, R. W. F., HOLSTEN, R. D., JACKSON, E. K. & BURNS, R. C. (1968). The acetylene-ethylene assay for nitrogen-fixation: laboratory and field evaluation. *Pl. Physiol.*, **43**, 1185.

HASKINS, E. F. & KIHARA, T. (1967). The use of spectrophotometry in an ecological investigation of the facultatively anaerobic purple photosynthetic bacteria. *Can. J. Microbiol.*, **13**, 1283.

HENDRIE, M. S., HODGKISS, W. & SHEWAN, J. M. (1970). The identification, taxonomy and classification of luminous bacteria. *J. gen. Microbiol.*, **64**, 151.

HUMM, H. J. (1946). Marine agar-digesting bacteria of the South Atlantic coast. *Duke Univ. Mar. Sta. Bull.*, **3**, 43.

JANNASCH, H. W. & MADDUX, W. S. (1967). A note on bacteriological sampling in seawater. *J. mar. Res.*, **25**, 185.

JOHNSON, P. T. (1968). A new medium for maintenance of marine bacteria. *J. Invert. Pathol.*, **11**, 144.

JONES, J. G. (1971). Studies on freshwater bacteria: factors which influence the population and its activity. *J. Ecol.*, **59**, 593.

JONES, J. G. (1972). Studies on freshwater bacteria: association with algae and alkaline phosphatase activity. *J. Ecol.*, **60**, 59.

KIMURA, T. (1961). A method for rapid detection of alginic acid digesting bacteria. *Bull. Fac. Fish., Hokkaido Univ.*, **12**, 41.

KING, E. O., WARD, M. K. & RANEY, D. E. (1954). Two simple media for the demonstration of pyocyanin and fluorescin. *J. Lab. clin. Med.*, **44**, 301.

KRISS, A. E. (1962). Suitability of Nansen water-sampler for microbiological investigations in seas and oceans. *Mikrobiologiya*, **31**, 1067. [In Russian, English translation *Microbiology*, **31**, 865 (1963).]

KRISS, A. E. (1963). *Marine Microbiology.* (Deep Sea.) Translated by J. M. Shewan and Z. Kabata. Edinburgh and London: Oliver & Boyd.

KRISS, A. E., LEBEDEVA, M. N. & TSIBAN, A. V. (1966). Comparative estimate of a Nansen and microbiological water bottle for sterile collection of water samples from depths of seas and oceans. *Deep-Sea Res.*, **13**, 205.

LISTON, J. (1956). Quantitative variations in the bacterial flora of flatfish. *J. gen. Microbiol.*, **15**, 305.

LISTON, J., HOLMAN, N. & MATCHES, J. (1969). Psychrophilic clostridia from marine sediments. *Bact. Proc.*, p. 35.

LOPONITSINA, V. V., NOVOZHILOVA, M. I. & KONDRAT'EVA, E. N. (1969). Photosynthetic bacteria isolated from the Aral Sea. *Mikrobiologiya*, **38**, 358.

LYMAN, J. & FLEMING, R. H. (1940). Composition of sea water. *J. mar. Res.*, **3**, 134.

MACAN, T. T. (1970). *Biological Studies of the English Lakes.* London: Longman (Publishers) Ltd.

MACKERETH, F. J. H. (1964). An improved galvanic cell for determination of oxygen concentration in fluids, *J. Scient. Instrum.*, **41**, 38.

MACLEOD, R. A. (1965). The question of the existence of specific marine bacteria. *Bact. Rev.*, **29**, 9.

MACLEOD, R. A., ONOFREY, E. & NORRIS, M. E. (1954). Nutrition and metabolism of marine bacteria. I. Survey of nutritional requirements. *J. Bact.*, **68**, 680.

Marine Biology IV (1968). Proceedings of the Fourth International Interdisciplinary Conference. Princeton, January 1966. *Unresolved Problems in Marine Microbiology* (C. H. Oppenheimer, ed.). New York: New York Academy of Sciences Interdisciplinary Communications Program.

MELAND, S. M. (1963). Marine alginate-decomposing bacteria from North Norway. *Nytt Magasin Botanik*, **10**, 53.

MORTIMER, C. H. (1940). An apparatus for obtaining water from different depths for bacteriological examination. *J. Hyg., Camb.*, **40**, 641.

NISKIN, S. J. (1962). A water sampler for microbiological studies. *Deep-Sea Res.*, **9**, 501.

PFENNIG, N. (1965). Anreicherungskulturen für rote und grüne Schwefelbakterien. In *Anreicherungskultur und Mutantenauslese*. Symposium in Gottingen, 28–30 April 1964. *Zentbl. Bakt. ParasitKde* Abyt. I, Supplementheft 1, p. 179. Stuttgart: Gustav Fischer Verlag.

PFENNIG, N. & LIPPERT, K. D. (1966). Über das Vitamin B_{12}—Bedürfnis phototropher Schwefelbakterien. *Arch. Mikrobiol.*, **55**, 245.

REPORT (1970). Of the *University College of Wales, Iceland Expedition*. In the Biology Library, University College of Wales, Aberystwyth, UK.

ROSS, S. S. (1963). A study of yeasts of marine origin. *Ph.D. Thesis*. Glasgow: University of Glasgow.

ROSS, S. S. & MORRIS E. O. (1964). An investigation of the yeast flora of marine fish from Scottish Coastal waters and a fishing ground off Iceland. *J. appl. Bact.*, **28**, 224.

SCHEGG, E. (1970). A new bacteriological sampling bottle. *Limnol. Oceanogr.*, **15**, 820.

SCHÖLLHORN, R. & BURRIS, R. H. (1966). Inhibition of nitrogen fixation by acetylene. *Fedn. Proc. Am. Socs exp. Biol.*, **25**, 710.

SHINANO, H. (1962). Studies on yeasts isolated from various areas of the North Pacific. *Bull. Jap. Soc. Sci. Fish.*, **28**, 1113. [In Japanese, English summary.]

SIEBURTH, J. McN., FREY, J. A. & CONOVER, J. T. (1963). Microbiological sampling with a piggy-back device during routine Nansen bottle casts. *Deep-Sea Res.*, **10**, 757.

SOROKIN, YU. I. (1960). Bathometer for taking samples of water for bacteriological analysis. *Byull. Inst. Biol. Vodokhran.*, No. 6, 53.

SOROKIN, YU. I. (1964). Question of a method of microbiological sampling in the sea in the light of recent developments in marine microbiology. *Okeanologiya*, **4**, 349. [In Russian.]

STEWART, W. D. P. (1968). Nitrogen input into aquatic ecosystems. In *Algae, Man and His Environment. Proc. Int. Symp., Syracuse*, June 1967 (D. F. Jackson, ed.) Syracuse: Syracuse University Press.

STEWART, W. D. P. (1969). Biological and ecological aspects of nitrogen fixation by free-living micro-organisms. *Proc. R. Soc. Lond. B.*, **172**, 367.

STEWART, W. D. P., FITZGERALD, G. P. & BURRIS, R. H. (1967). *In situ* studies on nitrogen fixation using the acetylene reduction technique. *Proc. natn. Acad. Sci. USA*, **58**, 2071.

TILTON, R. C., COBET, A. B. & JONES, G. E. (1967). Marine thiobacilli. I. Isolation and distribution. *Can. J. Microbiol.*, **13**, 1521.

TWEDT, R. M. & NOVELLI, R. M. E. (1971). Modified selective and differential isolation medium for *Vibrio parahaemolyticus*. *Appl. Microbiol.*, **22**, 593.

VAN UDEN, N. & CASTELO-BRANCO, R. (1963). Distribution and population densities of yeast species in Pacific water, air, animals and kelp off South California. *Limnol. Oceanogr.*, **8**, 323.

WEYLAND, H. (1969). Actinomycetes in North Sea and Atlantic Ocean sediments. *Nature, Lond.*, **223**, 858.

WILLIAMS, E. D. F. (1969). A submerged membrane filter apparatus for microbiological sampling. *Marine Biol.*, **3**, 78.

WINOGRADSKY, S. (1888). *Zur Morphologie und Physiologie der Schwefelbakterien,* p. 115. Leipzig: A. Felix.

ZOBELL, C. E. (1941*a*). Studies on marine bacteria. The cultural requirements of heterotrophic aerobes. *J. mar. Res.,* **4,** 42.

ZOBELL, C. E. (1941*b*). Apparatus for collecting water samples from different depths for bacteriological analysis. *J. mar. Res.,* **4,** 173.

ZOBELL, C. E. (1946). *Marine Microbiology.* Waltham Mass.: Chronica Botanica Co.

ZOBELL, C. E. & RITTENBERG, S. C. (1938). The occurrence and characteristics of chitinoclastic bacteria in the sea. *J. Bact.,* **35,** 275.

General Principles and Problems of Soil Sampling

S. T. WILLIAMS AND T. R. G. GRAY

Hartley Botanical Laboratories, P.O. Box 147, University of Liverpool, Liverpool L69 3BX, England

Before discussion of the principles and methods of sampling the microbial populations of soil, it is necessary to consider the purposes for which such samples are taken. There are obviously a great many reasons for attempting to detect or isolate soil microbes but two broad aims can be recognized. Firstly, the soil is often used solely as a source of microbes which are isolated and screened for useful attributes, such as production of antibiotics. Sampling presents no great problems and the means are justified by the end product. It might be argued, however, that recognition of the heterogeneity of microbial distribution in soil could result in detection of useful microbes overlooked when soil is treated as an homogenous material. Secondly, soil may be sampled to provide information on the natural distribution and activities of microbes. Samples taken must therefore be as representative as possible of the whole population under study and in their transport and storage, attempts are made to preserve their natural state.

General Principles of Sampling

When a soil is sampled, one or more of the attributes of its microbial population may be estimated. The precision of this estimate depends on the degree of error in the analytical procedures and on how representative the samples are of the whole population. In materials where a random distribution of microbes can be assumed, sampling error can be kept within reasonable limits but where distribution is not random, accurate sampling is difficult and results may be invalid (Cowell and Morisetti, 1969). In soil, microbial distribution is influenced by many factors, such as depth, roots, animals and water, and is therefore rarely random. The difficulty, if not impossibility, of drawing representative samples results in sampling errors which are almost always greater than those of the subsequent procedures. It is not usually economically feasible to draw sufficient samples to provide a highly accurate estimate. Therefore sampling procedures are essentially compromises between economy and accuracy. It is essential that sampling procedures are planned so that statistical analysis can be carried out to assess the inaccuracies. To achieve this, the following points should be borne in mind.

(a) In any study, a basic soil unit can be recognized. This can range from an acre field to a particular type of mineral grain but whatever its size, the results obtained are related to it. Comparisons are made between units of equivalent ranks and the variability of the samples taken to represent each individual unit can be assessed statistically.

(b) At least two and preferably more samples should be drawn from a unit and analysed separately to permit any assessment of variability. Therefore several small samples from a unit are preferable to one large one.

(c) The soil unit should be as homogenous as possible. If necessary, obviously heterogenous units can be subdivided into smaller, more homogenous ones. Thus, if two soils are being compared and each has two distinct horizons, these should be sampled and analysed separately as four units.

(d) The sampling area should be selected to be as homogenous as possible with respect to such factors as plant cover, soil type, topography, etc.

A number of basic sampling procedures, which meet some or all of these requirements are possible. Useful information on the statistical analysis of soil samples is given by Cline (1944), Peterson and Calvin (1965) and Parkinson, Gray and Williams (1971).

Simple random samples

These can be taken by selecting sampling points within a unit using pairs of random numbers from tables (Fisher and Yates, 1963) to select distances on co-ordinates of a grid system laid over the unit. Samples are taken where lines from the co-ordinates intersect.

Stratified random samples

These can be used when sub-units are discernible within a unit. Samples are drawn randomly, as above, from each sub-unit. The more subdivision made, the greater is the precision of the estimate but obviously the amount of work also increases.

Systematic sampling

Samples are taken at regular intervals in one or two dimensions. Parallel lines or grids are used to define sampling points. Care must be taken if regular fluctuations within the unit occur. Thus if there are rows of plants, highly selective sampling would occur if the sampling points coincided with the rows.

Sub-sampling

Sometimes a number of smaller sub-samples are taken from a sample after it has been drawn. This provides an estimate of the attributes of the large sample without dealing with it in its entirety. However, precision is reduced as an additional source of variation (that among sub-samples) is introduced.

Composite samples

Samples drawn separately from a single unit are sometimes then bulked and mixed. The whole mixture, or sub-samples from it, is then analysed. Sometimes this procedure is used to produce what is described as a "more representative" sample. However, this is not justified unless it has been proved by analysis of the variation among the original samples taken from the unit.

Collection and Storage of Samples

If soil is only to be screened for useful microbes, it will not generally be necessary to obtain undisturbed, representative samples. Therefore collection of such samples requires little beyond recognition of the gross units sampled (e.g. soil type, soil horizon) for future reference. Similarly the transport and storage of samples requires few precautions. It may even be feasible to keep soil for several months in the laboratory before it is used. However, storage conditions can induce changes in the microbial populations and under extreme conditions, potentially useful microbes may be lost. Thus some Gram negative bacteria die in air-dried soil (Stevenson, 1956; Jensen, 1962). A suitable method for dealing with such samples was described by Clark (1965). A large sample or mixed composite one is placed in a polythene bag or sealed waxed cardboard container. Care is taken not to expose it to undue heat or drying and it is stored at 4° until required.

When information on the natural state of the soil population is required, the choice of methods for the collection and subsequent transport of samples is more critical. As previously stated, sampling procedures should be designed to permit statistical evaluation of the results. Methods of taking the individual samples will depend very much on the type of soil unit under study. Many devices are used, including cellotape strips for the collection of individual particles (Mayfield, Williams, Ruddick and Hatfield,

1972), coring devices (e.g. glass tubes) for 5–10 g samples and complex mechanical augurs for taking large intact cores. Whatever the device used, some general precautions should be taken. Implements and containers should be sterilized as far as possible and care should be taken to avoid contamination of one unit by another. As far as possible, similar sized samples should be taken to permit accurate evaluation of the variation among them.

Ideally, the samples should be kept in the same physical, chemical and biological state as they were *in situ*. This can never be achieved but possible changes can be minimized, although there is no generally agreed procedure for dealing with samples. Temperatures in the laboratory will often differ considerably from those in the field and this represents one of the potential causes of change in the microbial populations. Some workers (Jensen, 1962) recommend that samples are kept refrigerated (0–5°), while others believe that this causes changes and store at room temperature (Casida, Klein and Santoro, 1964). Another possibility is to preserve the soil at the ambient sampling temperature in a thermos flask or similar container.

Changes in aeration and moisture content of samples are almost inevitable. To minimize these, a large block of soil may be taken to the laboratory where it is sub-sampled immediately before use. The feasibility of this procedure depends largely on the physical nature of the soil. Alternatively, normal sized samples can be taken and stored in polythene bags which allow relatively free passage of air but not moisture (Stotzky, Goos and Timonin, 1962). These may be sealed in another bag containing a few drops of water (Casida *et al.*, 1964).

An obvious way of reducing the effects of these and most other factors is to use the samples as quickly as possible. There is evidence of changes occurring within a few hours of sampling (James and Sutherland, 1939; Jensen, 1962) and therefore samples should be analysed within about 6 h (Jensen, 1968).

Many workers recommend sieving of samples before analysis, a 2 mm sieve often being chosen. This is done to obtain more reproducible results. In effect the natural heterogeneity of the soil is counteracted before analysis. This is justified when information on the natural distribution of microbes within the soil is not required. However, if sieving is used to produce composite samples, difficulties of statistical analysis as outlined above arise. It must also be appreciated that sieving is to some extent selective and microbes associated with the larger particles and aggregates will be excluded.

TABLE 1. The occurrence of organic substrates in a pine forest soil*

Substrates		% total internal surface area		% total organic surface area	
		F$_2$ layer	H layer	A$_1$ horizon	C horizon
Pine branches	cortex	0·004–0·037	0·005–0·017	0·016–0·17	0·009–0·076
	stele	0·001–0·019	0·002–0·010	0·003–0·07	0·002–0·034
	immature shoots	0·0–0·006	—	—	—
Pine dwarf shoots	cortex	4·6–7·1	—	—	—
	stele	2·2–6·3	—	—	—
Charcoal		0·0–0·02	0·0–0·04	0·0–0·93	0·0–0·9
Bark		0·001–0·023	0·0–0·005	—	—
Foliage leaves		61·3–71·2	—	—	—
Roots	live	—	0·0–0·12	1·0–18·9	1·97–29·0
	dead	—	0·0–0·09	5·3–24·3	7·0–30·3
	mycorrhizal	—	—	45·9–90·0	32·1–61·6
Cones	male	0·0–0·036	—	—	—
	female	0·03–0·20	—	—	—
	partly eaten	0·0–0·007	—	—	—
Seeds		0·0–0·17	—	—	—
Hyphae		0·09–0·18	0·02–0·08	0·27–2·3	0·4–3·0
Sclerotia		—	—	0·0–0·05	0·0–0·08
Resin aggs.		—	0·01–0·06	0·3–0·6	0·31–1·75
Insect exoskeletons		0·0–0·004	0·004–0·02	0·0–0·05	0·0–0·02
Faecal pellets		—	—	—	—
Unidentified		17·0–29·6	99·6–99·9	1·07–4·6	3·8–14·75

* After Hill (1967).

Detection of Microbes in Soil Units

The range of micro-environments within the soil is enormous. Even in an apparently simple developed soil such as that found on sand dunes colonized by pine trees, great differences exist within it. Thus in a study by Hill (1967), many distinct units were recognized. Tables 1 and 2 summarize

TABLE 2. The occurrence of various types of mineral grains in a pine forest soil*

Mineral	Frequency of occurrence (%)	
	A_1 horizon	C horizon
Quartz	60–73	63–70
Cryptocrystalline quartz	20–28	20–25
Cryptocrystalline quartz + iron oxide inclusions	5–11	4–15
Cryptocrystalline quartz + unoxidized iron inclusions	0·3 –0·4	0·21–0·39
Tormaline	0·01–0·02	0·0 –0·02
Glauconite	0·04–0·09	0·01–0·09
Pyroxene	0·10–0·19	0·10–0·20
Aluminosilicate	0·10–0·15	0·10–0·20
Corundum	0·01–0·03	0·05–0·09
Cordierite	0·02–0·04	0·02–0·04
Calcite	0·0 –0·01	0·01–0·19
Calcite + iron	0·07–0·13	0·31–0·56
Calcite + white inclusions	0·01–0·02	0·0 –0·01
Scale	0·0 –0·02	0·0 –0·03
Unidentified	0·16–0·35	0·26–0·38

* After Hill (1967).

the principal organic and mineral substrates found in the A_o, A_1 and C horizons of this soil. It is clear that while leaves are characteristic of the organic matter in the A_o layer, roots make up most of the organic material in the A_1 and C horizons. The occurrence of minerals is more constant though calcite is more abundant in the alkaline C horizon. It was also noted that many of the mineral particles had a thin coating of organic matter.

Any of these materials can be considered as a soil unit and may be inhabited by particular microbes. Thus Gray and Baxby (1968) showed that *Sporothrix* sp, *Cylindrocarpon radicicola*, and *Cephalosporium* sp were commonly found on hyphae in the soil, while *Verticillium* sp and *Mortierella marburgensis* were associated with insect exoskeletons. Parkinson and

Balasooriya (1967) found that *Zygorhynchus moelleri, Gliomastix guttuli-formis* and *Chaetosphaeria myriocarpa* were restricted to mineral particles and, amongst other fungi, *Phoma* sp and *Paecilomyces carneus* occurred only on unidentifiable organic fragments. Other workers have shown a clear preference of certain organisms for the root surface (Macura, 1968) and bacteria able to utilize glucose, acetate, alanine, as well as ammonifiers, denitrifiers, starch hydrolyzers and resazurin-reducing bacteria are all relatively more abundant in the root zone. Recognition of soil units is therefore of prime importance before sampling is undertaken.

Methods for Detection of Microbes in Various Sized Soil Units

Large soil units

Often it is only a general survey of the microbial populations that is required. In this case it is unnecessary to sample small soil units and relatively large, heterogenous units, such as a gram of soil, are examined. Samples may be taken from a variety of soil types, horizons or differently treated soils. The following general methods are all useful for such an investigation.

Dilution plate procedures

These have been used to count numbers of cells of bacteria or spores of actinomycetes and fungi in soil. They can also be used in general isolation programmes. For isolation purposes, precise dilutions are unnecessary and the pour plate method is adequate (Parkinson *et al.*, 1971). Surface plating (Ledeberg and Ledeberg, 1952) may be useful since it can be combined with replica plating to test the responses of the population to different nutritional and environmental conditions. Media used depend on the aims of the study and include those given on p. 119.

Soil plate method (Warcup, 1950)

Sometimes the number of cells of an organism in soil may be so low that dilution of the sample is unnecessary. This is frequently so for fungi. Between 0·005–0·015 g of soil (a soil crumb) is dispersed in molten agar in a Petri dish. After incubation for 2 days, the plates are observed daily for fungal growth and the initial, faster growing colonies subcultured to prevent their overgrowth of slower growing ones. As with dilution procedures, the fungi isolated are often present in the soil as spores but the simplicity of the method makes it valuable for large scale-general surveys.

Small soil units

Sieving techniques

Preliminary sieving of soil facilitates the selection of different sized units. Frequently, roots and decomposing leaf material are larger than the mineral fraction of soil. Sieved material may be examined by eye or by microscopy and various soil units separated. Mineral grains may be identified under white or ultra-violet light and organic materials can often be recognized by their shape and surface structure. The selected units can then be plated as described for the larger units.

Washing techniques

These have been designed to separate soil into its constituent units and to remove loosely adhering microbial propagules, leaving only those intimately associated with them. This increases the chances of isolating those present as hyphae, slower growing fungi and microbes situated inside particles. Bacteria are too readily detached to be isolated efficiently but actinomycetes can be, provided that faster growing fungi are suppressed (see below). The most widely used washing techniques for roots and leaf litter are those of Harley and Waid (1955) and for soil mineral and organic particles those of Williams, Parkinson and Burges (1965).

In the Harley and Waid method, the material is placed in a sterile screw-topped vial containing 10 ml of sterile water. The vial is agitated mechanically for 2 min and the water decanted. This procedure is repeated until only a low, constant number of propagules are removed as determined by plating of the washing water. The washed material is then transferred aseptically to a sterile filter paper, dried and plated on an appropriate medium. While the method is useful for larger substrates such as roots and leaves, it is not easily applied to smaller particles of low specific gravity which are removed in the washing water.

In the second method, samples are placed in alcohol sterilized Perspex boxes fitted with a number of graded sieves. Soil is placed on the upper, largest mesh sieve and sterile water added. The mixture is agitated by passage of sterile air and after 2 min the water is drained off and the procedure repeated. On completion of washing, soil particles are distributed on the different sized sieves and can be removed and plated according to type and size. Difficulties arise when the soil contains much clay or humus as the washing is less efficient. Building the apparatus involves some effort and unless an extensive study of soil fungi is envisaged, the simpler method of Harley and Waid (1955) or direct soil plating will be sufficient.

If it is necessary to study the internal colonizers of soil organic matter, then washing may be combined with surface sterilization and subsequent fragmentation of the particles (Parkinson *et al.*, 1971).

General Isolation Media

It would be impossible to list all the useful isolation media used in general studies of soil organisms but some generalizations can be made.

Media can be made selective for fungi, bacteria or actinomycetes by alterations to their pH or by addition of selective inhibitors. pH values of 5·0–5·5 are suitable for fungi while higher values are required by most bacteria and actinomycetes. Nevertheless the existence of acidophilic bacteria and actinomycetes or basophilic fungi must be considered.

Fungal development may be suppressed by addition of either or both actidione (Light) and nystatin (Squibb) at concentrations of 50 μg/ml of medium. Bacteria can be suppressed by addition of aureomycin (Cyanamid) at 30 μg/ml. Most Gram positive bacteria can be suppressed by adding 1 to 10 μg/ml penicillin (Glaxo) and Gram negative forms by adding 5 to 10 μg/ml polymixin B sulphate (Burroughs Wellcome). Media which can be used include the following:

Soil extract agar (for fungi, bacteria and actinomycetes)

A suitable procedure for making the extract was described by James (1958). One kg of soil is autoclaved with 1 litre of water for 20 min at 20 lb/in^2. The liquid is strained and made up to 1 litre. If it is cloudy, a little calcium sulphate is added and after being allowed to stand, it is filtered. The extract may be sterilized and solidified with agar as it is or added to a medium with other nutrients incorporated.

Peptone yeast extract agar (Goodfellow, Hill and Gray, 1968; for heterotrophic bacteria)

Peptone, 5·0 g; yeast extract, 1 g; ferric phosphate, 0·01 g; agar; distilled water, 1 litre, pH 7·2.

Starch casein agar (Küster and Williams, 1964; for actinomycetes)

Starch 10·0 g; casein (vitamin free), 0·3 g; KNO_3, 2·0 g; NaCl, 2·0 g; K_2HPO_4, 2·0 g; $MgSO_4.7H_2O$, 0·05 g; $CaCO_3$, 0·02 g; $FeSO_4.7H_2O$, 0·01 g; agar; distilled water, 1 litre, pH 7·2.

Czapek-Dox agar (for fungi)

Sucrose, 30·0 g; NaNO₃, 2·0 g; K₂HPO₄, 1·0 g; KCl, 0·5 g; MgSO₄.7H₂O, 0·5 g; FeSO₄, trace; agar; distilled water, 1 litre. Adjust pH to 5·5. Yeast extract (6·5 g) may also be added.

Other general media and those for specific groups of soil microbes are given by Parkinson *et al.* (1971).

References

CASIDA, L. E., KLEIN, D. A. & SANTORO, T. (1964). Soil dehydrogenase activity. *Soil Sci.*, **98**, 371.

CLARK, F. E. (1965). Agar plate method for total microbial count. In *Methods of Soil Analysis*. II (C. A. Black *et al.*, eds). pp. 1460–66. Madison, USA: American Soc. of Agronomy.

CLINE, M. G. (1944). Principles of soil sampling. *Soil Sci.*, **58**, 275.

COWELL, N. P. & MORISETTI, M. D. (1969). Microbiological techniques—some statistical aspects. *J. Sci. Fd Agric.*, **20**, 573.

FISHER, R. A. & YATES, F. (1963). *Statistical Tables for Biological, Agricultural and Medical Research* 6th. Ed. Edinburgh: Oliver & Boyd.

GOODFELLOW, M., HILL, I. R. & GRAY, T. R. G. (1968). Bacteria in a pine forest soil. In *The Ecology of Soil Bacteria*, (T. R. G. Gray and D. Parkinson, eds). pp. 500–15. Liverpool: Liverpool University Press.

GRAY, T. R. G. & BAXBY, P. (1968). Chitin decomposition in soil. II. The ecology of chitinoclastic micro-organisms in forest soil. *Trans. Br. mycol. Soc.*, **51**, 293.

HARLEY, J. L. & WAID, J. (1955). A method for studying active mycelia on living roots and other surfaces in the soil. *Trans. Br. mycol. Soc.*, **38**, 104.

HILL, I. R. (1967). *Application of the fluorescent antibody technique to an ecological study of bacilli in soil*. Ph.D. Thesis. Liverpool: University of Liverpool.

JAMES, N. (1958). Soil extract in soil microbiology. *Can. J. Microbiol.*, **4**, 363.

JAMES, N. & SUTHERLAND, M. (1939). The accuracy of the plating method for estimating the number of bacteria and fungi from one dilution and one aliquot of a laboratory sample of soil. *Can. J. Res.*, **17**, 97.

JENSEN, V. (1962). Studies on the microflora of Danish beech forest soils. I. The dilution plate count technique for the enumeration of bacteria and fungi in soil. *Zentbl. Bakt. ParasitKde. Abt.* II. **116**, 13.

JENSEN, V. (1968). The plate count technique. In *The Ecology of Soil Bacteria* (T. R. G. Gray and D. Parkinson, eds). pp. 158–70. Liverpool: Liverpool University Press.

KÜSTER, E. & WILLIAMS, S. T. (1964). Selection of media for isolation of streptomycetes. *Nature, Lond.*, **202**, 928.

LEDEBERG, J. & LEDEBERG, E. M. (1952). Replica plating and indirect selection of bacterial mutants. *J. Bact.*, **63**, 399.

MACURA, J. (1968). Physiological studies of rhizosphere bacteria. In *The Ecology of Soil Bacteria* (T. R. G. Gray and D. Parkinson, eds). pp. 379–95. Liverpool: Liverpool University Press.

MAYFIELD, C. I., WILLIAMS, S. T., RUDDICK, S. M. & HATFIELD, H. L. (1972). Studies on the ecology of actinomycetes in soil. IV. Observations on the form and growth of actinomycetes in soil. *Soil Biol. Biochem.*, **4**, 79.

PARKINSON, D. & BALASOORIYA, I. A. (1967). Fungi in a pine wood soil. I. *Rev. d'Ecol. Biol. Sol*, **4**, 463.

PARKINSON, D., GRAY, T. R. G. & WILLIAMS, S. T. (1971). *Methods for studying the Ecology of Soil Micro-organisms.* Oxford and Edinburgh: Blackwell.

PETERSON, R. G. & CALVIN, L. D. (1965). Sampling. In *Methods of soil analysis* I. (C. A. Black *et al.* eds). pp. 54–72. Madison: American Soc. of Agronomy.

STEVENSON, I. L. (1956). Some observations on the microbial activity in re-moistened air-dried soils. *Pl. Soils*, **8**, 170.

STOTZKY, G., GOOS, R. D. & TIMONINN, M. I. (1962). Microbial changes occurring in soil as a result of storage. *Pl. Soils*, **16**, 1.

WARCUP, J. H. (1950). The soil plate method for isolation of fungi from soil. *Nature, Lond.*, **166**, 117.

WILLIAMS, S. T., PARKINSON, D. & BURGESS, N. A. (1965). An examination of the soil washing technique by its application to several soils. *Pl. Soil*, **22**, 167.

Methods for the Morphological Examination of Aerobic Coryneform Bacteria

G. L. CURE AND R. M. KEDDIE

Department of Microbiology, The University,
Reading RG1 5AQ, England

Coryneform bacteria are numerically important in a wide range of habitats including soil (see Keddie, Leask and Grainger, 1966) and their recognition depends entirely on microscopical appearance and staining reactions. However not only does the morphology change during the growth cycle but it is also markedly influenced by the cultural conditions, and in particular by the composition of the medium used. It is therefore unfortunate that many different media have been used for the morphological study of different coryneform bacteria. It would be extremely useful for comparative purposes if a single, reproducible medium could be used for aerobic coryneform bacteria from a wide range of habitats. In the following pages we describe the medium and methods we use for morphological examinations. As far as is practicable a standard medium and method are used for strains from different habitats: at fixed times throughout the growth cycle microscopical preparations are made and photographed so that direct comparisons between strains can later be made. The photomicrographs are also used for measurements of cell size.

The Coryneform Group: a Definition

With Jensen's (1952) description as a guide we may define the coryneform group in the following way. In exponential phase cultures in complex media, irregular rods occur which vary considerably in size and shape and include straight, bent and curved, wedge-shaped and club-shaped forms. A proportion of the rods are arranged at an angle to each other to give V-formations but other angular arrangements may be seen. Rudimentary branching may occur, especially in richer media, but definite mycelia are not formed. In stationary phase cultures the cells are generally much shorter and less irregular and a variable proportion is coccoid in shape. The rods may be non-motile or motile; endospores are not formed. They are

Gram positive but may be readily decolorized and may show only Gram positive granules in otherwise Gram negative cells. They are not acid-fast.

Arthrobacter: *morphological changes during the growth cycle*

Within the coryneform group, members of the genus *Arthrobacter* show the most marked changes in morphology during the growth cycle and the sequence of morphological changes that occurs is a major distinguishing feature of the genus (Fig. 1). Older cultures, usually 2–7 days, are composed entirely, or largely, of coccoid cells which in some strains may resemble micrococci, especially in stained preparations. When coccoid cells are transferred to fresh complex medium, they swell slightly and then produce outgrowths from one or occasionally more parts of the cell. The growth and division which follow give rise to the irregular rods (described above) characteristic of exponential phase cultures. As growth proceeds the rods become shorter and are eventually replaced by the coccoid cells characteristic of stationary phase cultures. The coccoid cells are formed either by a gradual shortening of the rods at each successive division or, especially in richer media, by multiple fragmentation of larger rods.

The Effect of Medium on Morphology

The extent to which the morphology changes during the growth cycle is markedly dependent on the growth medium. This effect has been most studied in members of the genus *Arthrobacter*. At one extreme, some species have been shown to multiply only in the coccoid form in simple, glucose mineral salts media (Ensign and Wolfe, 1964; Veldkamp, van den Berg and Zevenhuizen, 1963). At the other extreme, growth of *A. globiformis* in media containing high concentrations of complex organic constituents results in marked elongation of the rods and rudimentary branching is prominent; final transformation into coccoid forms may be considerably delayed (Stevenson, 1961; Veldkamp *et al.*, 1963). A rather similar but even more pronounced effect of medium on the morphology of *A. ureafaciens* was attributed to a deficiency of magnesium ions in the complex medium used (Blankenship and Doetsch, 1961). In some arthrobacters at least, growth in media with a high carbon/nitrogen ratio results in the development of entire populations of large coccoid cells sometimes referred to as "cystites" (Stevenson, 1963; Mulder *et al.*, 1966).

Choice of Medium

The medium used for morphological studies must therefore be chosen with care: media containing low concentrations of complex organic nutrients

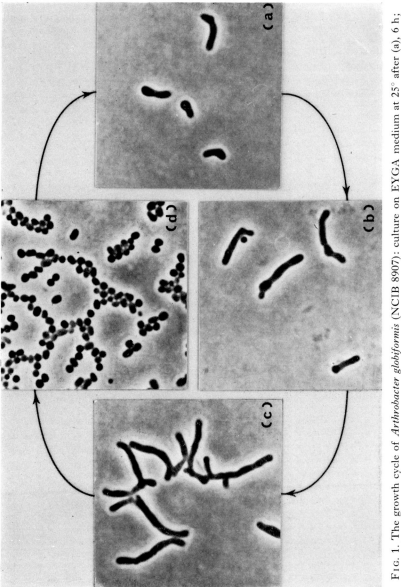

FIG. 1. The growth cycle of *Arthrobacter globiformis* (NCIB 8907): culture on EYGA medium at 25° after (a), 6 h; (b), 12 h; (c), 24 h; (d), 3 days. × 1875.

but nutritionally adequate with respect to both mineral nutrients and organic growth factors are probably the most generally satisfactory. Those media based on soil extract which have been used for soil coryneform bacteria may fulfil most of these requirements e.g. YS medium (Lochhead and Burton, 1955). However such media tend to give inconsistent results in morphological studies because of the variable nature of soil extract and because some mineral components are precipitated to varying extents during preparation.

With the above considerations in mind, we have devised a medium,

TABLE 1. The preparation of Mineral Base E (Owens and Keddie, 1969)

Stock solutions (400 ml quantities in glass-distilled water):

I		K_2HPO_4	80·0 g
II		KH_2PO_4	62·4 g
III		$CaCl_2$	2·0 g
IV	}	$MgSO_4.7H_2O$	8·0 g
		NaCl	4·0 g
V		$(NH_4)_2SO_4$	20·0 g

EDTA/Trace metals mix. To 600 ml glass-distilled water add:

EDTA	5·0 g
$ZnSO_4.7H_2O$	2·2 g
$MnSO_4.4H_2O$	0·57 g
$FeSO_4.7H_2O$	0·50 g
$CoCl_2.6H_2O$	0·161 g
$CuSO_4.5H_2O$	0·157 g
$Na_2MoO_4.2H_2O$	0·151 g

Adjust to pH 6·0 by adding small quantities of 40% KOH (the EDTA will go into solution on addition of alkali) then make up to 1000 ml with glass-distilled water. Stock solutions are stored at $-20°C$ to prevent microbial growth.

To prepare 1 litre of Mineral Base E add the solutions to glass-distilled water in the order given:

Glass-distilled water	962·0 ml
Solution I	5·5 ml
Solution II	4·5 ml
Solution III	5·0 ml
Solution IV	10·0 ml
Solution V	10·0 ml
EDTA/Trace metals mix	3·0 ml

Final pH 6·8: the pH may be adjusted within the limits of the buffer system by altering the relative amounts of solutions I and II provided that a total of 10 ml/litre is added.

Mineral Base E-N is prepared in the same way as Mineral Base E but contains no $(NH_4)_2SO_4$.

Sterilize at 121° for 20 min for quantities up to 100 ml.

EYGA, which should be reproducible and which can be used for morphological studies on coryneform bacteria from widely different habitats. The medium is based on the chelated Mineral Base E developed by Owens and Keddie (1969) for nutritional studies on coryneform bacteria. When Mineral Base E is prepared as directed (Table 1), a clear solution is obtained and reproducibility is ensured: a precipitate which forms on heating completely redissolves on cooling. There is now considerable evidence that yeast extract supplemented with vitamin B_{12} will supply all the necessary organic growth factors for a large majority of coryneform bacteria from different sources (Keddie et al., 1966; Mulder et al., 1966; Skerman and Jayne-Williams, 1966; Owens and Keddie, 1968). These constituents, together with glucose form the basis of EYGA (Table 2).

TABLE 2. EYGA medium: composition and preparation

Composition

	Mineral Base E	1000 ml
	Vitamin B_{12}	2·0 μg
	Yeast extract (Difco)	1·0 g
	Agar (Oxoid No. 1)	12·0 g
	Glucose	1·0 g

Preparation

Add the yeast extract and agar to 1 litre of Mineral Base E (Table 1) and steam until the agar has dissolved. Dissolve the glucose and add 1 ml of vitamin B_{12} solution (2 μg/ml). The vitamin B_{12} solution is conveniently prepared by adding the contents of a 250 μg, 1 ml ampoule of vitamin B_{12} ("Cytamen", Glaxo Laboratories Ltd., Greenford, Middlesex) to 124 ml of glass-distilled water. Adjust the medium to pH 6·8 if necessary, dispense as required and sterilize at 121° for 20 min.

However strains are known which require unsaturated fatty acids e.g. *Corynebacterium bovis* (Skerman and Jayne-Williams, 1966) or which require sideramines such as the *terregens* factor (Burton, 1957).

In our work relatively few strains with these additional requirements have been encountered and therefore we prefer to supplement EYGA only when necessary. A requirement for unsaturated fatty acids may be satisfied by the addition to EYGA of Tween 80 (0·05% v/v). *Terregens* factor is most easily provided as *A. pascens* (NCIB 8910) culture filtrate (0·1% v/v) prepared as described by Lochhead and Burton (1953). However it should be noted that for some rarely isolated soil organisms, soil extract cannot be replaced by yeast extract + vitamin B_{12} + *terregens* factor (Grainger and Keddie, 1963).

Comparisons of the morphology of some commonly encountered coryneform bacteria when grown on EYGA and on an agar medium similar to the YS medium of Lochhead and Burton (1955), may be made by reference to Figs 2 and 3.

Methods

Incubation

The method described is used for coryneform bacteria in which the mode of metabolism is primarily or entirely respiratory and cultures are incubated in air, those of animal parasites at 37° (in a moist atmosphere) and all others at 25°.

Staining methods

Gram reaction

Heat-fixed smears are prepared from EYGA slope cultures after incubation for 18 h, 3 and 7 days and stained by a method similar to Hucker's modification of the Gram staining method (Cunningham, 1947, p. 60). Flood smear for 1 min with a mixture of crystal violet solution (2% w/v in absolute ethanol) one part, and ammonium oxalate solution (1% w/v aqueous) 4 parts. Pour off the stain and flood (removing all precipitate) for 1 min with iodine solution (1 g iodine and 2 g potassium iodide in 300 ml distilled water). Drain and wash off iodine solution with ethanol (*c.* 95%) and treat with fresh ethanol just to the point when stain is no longer visibly removed from the smear. Immediately wash thoroughly with water and then stain with a suitable counterstain.

Acid-fastness

This is tested for by preparing smears from 5 day slope cultures on TSXA containing 7% (w/v) glycerol (Keddie *et al.*, 1966) and staining by the method of Gordon and Smith (1953).

Methods for microscopical examination

In the method described living cultures are examined and photographed on thin layers of agar using phase-contrast microscopy. If care is exercised in making slide preparations there is minimal disturbance to cell arrangements; the use of agar layers overcomes the problem of Brownian movement.

Slide preparations

Pour *c.* 6 ml of molten EYGA (at *c.* 55°) into a 9 cm plastic Petri dish, distribute uniformly and allow to solidify on a *level* surface. Surface dry the agar plate (open and inverted) at 40° for 15 min, inoculate with one

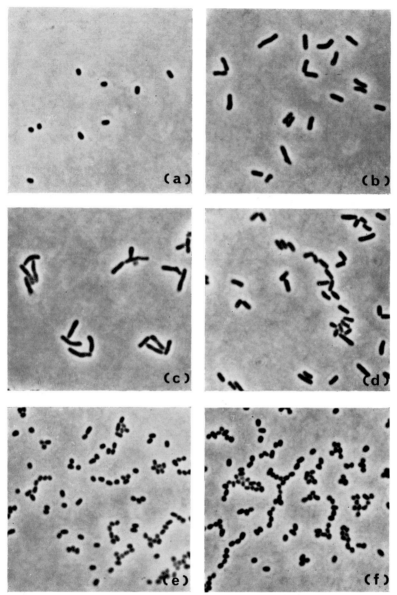

FIG. 4. Coryneform isolate from pig slurry when grown on EYGA medium at 25° after
(a), 0 h; (b), 6 h; (c), 12 h; (d), 24 h; (e), 3 days; (f), 7 days. × 1875.

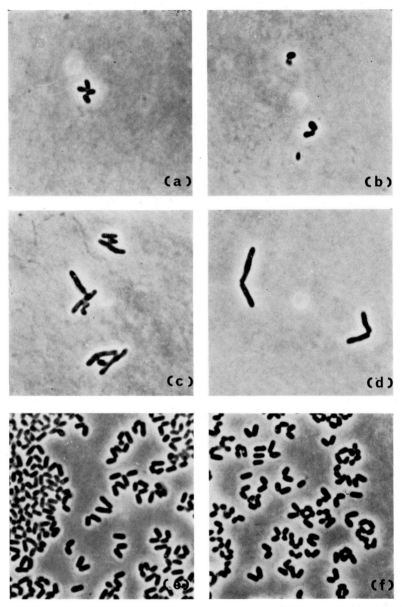

FIG. 5. Coryneform isolate from cauliflower when grown on EYGA medium at 25° after (a), 0 h; (b), 6 h; (c), 12 h; (d), 24 h; (e), 3 days; (f), 7 days. × 1875.

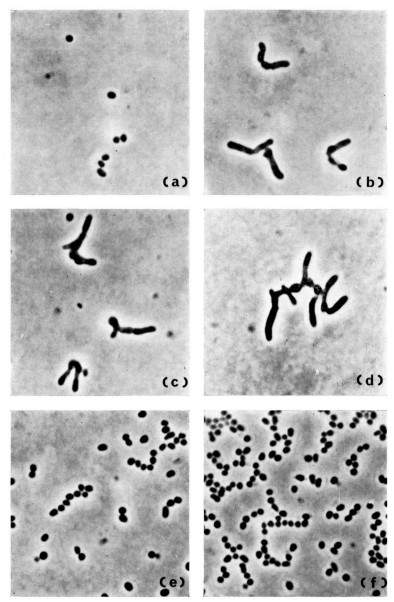

FIG. 6. The growth cycle of *Corynebacterium ilicis* (ATCC 14264), a plant pathogen: culture on EYGA medium at 25° after (a), 0 h; (b), 6 h; (c), 12 h; (d), 24 h; (e), 3 days; (f), 7 days. × 1875. Notice the morphological similarity to *Arthrobacter globiformis* (Fig. 1).

drop by Pasteur pipette of a just visibly turbid suspension of cells and spread the inoculum uniformly and gently over the plate with a previously sterilized bent glass spreader. Cell suspensions of uniform turbidity are prepared by suspending material from a 7 day EYGA slope in 3 ml sterile Mineral Base E-N (Table 1) and mixing by means of a "Rota-mixer" (Hook and Tucker Ltd., 301 Brixton Road, London SW9) or its equivalent. Slide preparations are made after incubation for 0, 6, 12 and 24 h, 3 and 7 days as follows: a piece of agar *c*. 1 cm^2 is cut and removed from the plate using a sterile scalpel and carefully placed on a clean slide; a No. 1 glass coverslip is then gently placed on the agar square and the edges sealed with rubber solution to prevent evaporation.

Photomicrography

Representative fields are photographed using Ilford Pan F 35 mm film. Development in Ilford ID-11 developer for 10 min at 20° gives negatives of suitable contrast. Prints are prepared at a standard, known magnification so that cells may later be measured on the prints using a Polaron × 10 lens with measuring graticule PS No. 8 (Polaron Equipment Ltd., 4 Shakespeare Road, London N3). Figs 4–6 show photomicrographs of coryneform isolates from different habitats prepared using the methods described.

The optical equipment used was a Wild M20 microscope with a Phase-Fluotar, × 100, oil immersion objective (N.A. 1·30) and × 10 compensating photo-eyepiece. Illumination was by quartz-iodine lamp in a Wild Universal Lamp-housing. A Wild Photoautomat with Mark a4 camera was used for photomicrography.

Acknowledgement

G. L. Cure is indebted to the Science Research Council for financial assistance during the course of this work.

References

BLANKENSHIP, L. C. & DOETSCH, R. N. (1961). Influence of a bacterial cell extract upon the morphogenesis of *Arthrobacter ureafaciens*. *J. Bact.*, **82**, 882.

BURTON, M. O. (1957). Characteristics of bacteria requiring the *terregens* factor. *Canad. J. Microbiol.*, **3**, 107.

CUNNINGHAM, A. (1947). *Practical Bacteriology*. 3rd Ed. Edinburgh: Oliver & Boyd.

ENSIGN, J. C. & WOLFE, R. S. (1964). Nutritional control of morphogenesis in *Arthrobacter crystallopoietes*. *J. Bact.*, **87**, 924.

GORDON, R. E. & SMITH, M. M. (1953). Rapidly growing, acid fast bacteria. I. Species' descriptions of *Mycobacterium phlei* Lehmann & Neumann and *Mycobacterium smegmatis* (Trevisan) Lehmann & Neumann. *J. Bact.*, **66**, 41.

GRAINGER, J. M. & KEDDIE, R. M. (1963). Nutritional studies on coryneform bacteria from soil and herbage. *J. gen. Microbiol.*, **31**, viii.

JENSEN, H. L. (1952). The coryneform bacteria. *A. Rev. Microbiol.*, **6**, 77.

KEDDIE, R. M., LEASK, B. G. S. & GRAINGER, J. M. (1966). A comparison of coryneform bacteria from soil and herbage: cell wall composition and nutrition. *J. appl. Bact.*, **29**, 17.

LOCHHEAD, A. G. & BURTON, M. O. (1953). An essential bacterial growth factor produced by microbial synthesis. *Canad. J. Bot.*, **31**, 7.

LOCHHEAD, A. G. & BURTON, M. O. (1955). Quantitative studies of soil micro-organisms. XII. Characteristics of vitamin B_{12}–requiring bacteria. *Canad. J. Microbiol.*, **1**, 319.

MULDER, E. G., ADAMSE, A. D., ANTHEUNISSE, J., DEINEMA, M. H., WOLDENDORP, J. W. & ZEVENHUIZEN, L. P. T. M. (1966). The relationship between *Brevibacterium linens* and bacteria of the genus *Arthrobacter*. *J. appl. Bact.*, **29**, 44.

OWENS, J. D. & KEDDIE, R. M. (1968). A note on the vitamin requirements of some coryneform bacteria from soil and herbage. *J. appl. Bact.*, **31**, 344.

OWENS, J. D. & KEDDIE, R. M. (1969). The nitrogen nutrition of soil and herbage coryneform bacteria. *J. appl. Bact.*, **32**, 338.

SKERMAN, T. M. & JAYNE-WILLIAMS, D. J. (1966). Nutrition of coryneform bacteria from milk and dairy sources. *J. appl. Bact.*, **29**, 167.

STEVENSON, I. L. (1961). Growth studies on *Arthrobacter globiformis*. *Canad. J. Microbiol.*, **7**, 569.

STEVENSON, I. L. (1963). Some observations on the so-called "cystites" of the genus *Arthrobacter*. *Canad. J. Microbiol.*, **9**, 467.

VELDKAMP, H., BERG, G. VAN DEN & ZEVENHUIZEN, L. P. T. M. (1963). Glutamic acid production by *Arthrobacter globiformis*. *Antonie van Leeuwenhoek*, **29**, 35.

The Isolation and Identification of Certain Soil Gram Negative Bacteria

A. J. HOLDING

Department of Microbiology, School of Agriculture,
University of Edinburgh, Edinburgh EH9 3JG, Scotland

Gram negative bacteria may be a predominant component of the microbial population in some soils and therefore their colonies may occur on dilution plates of non-selective media such as soil extract agar. However, since they frequently comprise a very small proportion of the population, selective procedures are also required for their isolation. This paper is concerned with the selective methods that are available for isolating the dominant heterotrophic aerobic and facultatively anaerobic bacteria, but no consideration is given to the isolation of autotrophic bacteria, nitrogen-fixing bacteria or other specialized groups of Gram negative organism. A brief description of a scheme available for the preliminary differentiation of the dominant heterotrophic group is also provided. The organisms being discussed are distributed in soil macroenvironments and microenvironments in close association with a wide range of other organisms and therefore there is no specific sampling procedure to obtain these bacteria. The general methods recommended for obtaining soil samples for the isolation of soil bacteria are used and these have been discussed elsewhere in this volume —see p. 111.

Isolation Procedures

Soil samples mechanically macerated and then diluted in 0·01% (w/v) peptone water have usually yielded the highest count of colonies of Gram negative bacteria.

The following procedures have been published for the isolation of various groups:

Predominant Gram negative bacteria

A medium recommended by Holding (1960) had the following composition (g/l): peptone (Difco), 5·0; meat extract (Lab Lemco) 5·0; agar (Davis),

15·0; pH 6·8. Sterile crystal violet solution (final concentration, 2 ppm) was added to the medium immediately prior to pouring the agar. After drying the plates, 0·1 ml of the appropriate dilution was spread evenly over the agar surface. This procedure enabled the colonies of organisms of several genera which developed to be differentiated in terms of colour, size and texture. Non-motile strains were rarely isolated and although they may not be a predominant type, further evidence is required to determine whether soil *Acinetobacter* strains are inhibited.

Saprophytic fluorescent pseudomonads

Sands and Rovira (1970) used a similar surface-plating procedure to that described above. The medium had the following composition (g/l): Protoese peptone No. 3 (Difco), 20·0; K_2SO_4, 1·5; $MgSO_4 7H_2O$, 1·5; Ionagar No. 1 (Oxoid), 12·0; glycerol (ml/l), 8·0; pH 7·2. Before pouring the agar, penicillin G (75 units/ml) novobiocin (45 μg/ml) and cycloheximide (75 μg/ml) were added.

Rhizobium *spp*

The medium described by Graham (1969) contained the following ingredients (g/l): mannitol, 5·0; lactose, 5·0; K_2HPO_4, 0·5; NaCl, 0·2; $CaCl_2$ $2H_2O$, 0·2; $MgSO_4.7H_2O$, 0·1, $FeCl_3.6H_2O$, 0·1; yeast extract, 0·5, and agar 20·0. After autoclaving, the following compounds were added (mg/l): cycloheximide, 200; pentachloronitrobenzene, 100; sodium benzyl penicillin, 25; chloromycetin, 10; sulfathiazole, 25; neomycin, 2·5 and, if desired, 2·5 ml of a 1% (w/v) solution of Congo red. Final pH should be adjusted to 7·0. Rapid advances in our understanding of the ecology of rhizobia in soil will be possible if this medium proves to be reliable for the isolation of rhizobia from very diverse soil types.

Agrobacterium *spp*

Antibiotics have also been incorporated as inhibitors in media for the isolation of strains of this genus. The medium recommended by Schroth, Thompson and Hildebrand (1965) for the isolation of *A. tumefaciens* and *A. radiobacter* had the following composition (g/l): mannitol, 10·0; $NaNO_3$, 4·0; $MgCl_2$, 2·0; $Mg(PO_4)_2$, 0·2; $MgSO_4.7H_2O$, 0·1; Ca propionate, 1·2; $NaHCO_3$, 0·075; $MgCO_3$, 0·075, and agar, 20·0. After autoclaving the medium and cooling to 50–55°, the following materials were added (ppm): Berberine, 275; Na selenite, 100; penicillin G (1625 units/mg), 60; streptomycin sulphate (78·1% streptomycin base), 30; cycloheximide (85–100% active ingredient), 250; tyrothricin (pure) 1.0; and bacitracin (65 units/mg), 100. The pH was adjusted to 7·1.

A medium selective for one group of strains of *A. tumefaciens* has been described by New and Kerr (1971). This highly selective medium contained (g/l): erythritol, 5·0; NaNO₃, 2·5; KH_2PO_4, 0·1; $CaCl_2$, 0·2; NaCl, 0·2; $MgSO_4.7H_2O$, 0·2; Fe EDTA solution (0·65%), 2 ml/l; biotin, 2 μg. After autoclaving and cooling, the following inhibitors were added (ppm): cyclo-heximide, 250; bacitracin, 100; tyrothricin 1·0 and Na selenite, 100.

Clark (1969) found that high concentrations of $MnSO_4.4H_2O$ in a lactose mineral salts medium were very selective for agrobacteria.

Acinetobacter *spp*

Although no direct plating procedure has been recommended, Baumann (1968) reported that these organisms could be isolated from soil by enrich-ment in an acetate-nitrate liquid medium followed by plating on to the same medium or a yeast extract agar. The liquid medium contained the following (g/l): Na acetate, 2·0; KNO₃, 2·0; $MgSO_4.7H_2O$, 0·2; 0·04 M–KH_2PO_4–Na_2HPO_4 buffer, trace elements, pH 6·0.

Preliminary Identification

Observations of 4 characteristics enable this differentiation, mainly to the generic level, to be undertaken.

(a) Production of a pigment.

(b) Ability to utilize glucose—if positive, whether fermentative and/or oxidative breakdown.

(c) Motility—if positive, whether due to flagella or gliding movement.

(d) Arrangement of flagella, if present.

The identification using these criteria is summarized in Table 1.

TABLE 1. Identification of common Gram negative bacteria

Pigment produced	Glucose utilization	Motility	Position of flagella	Genus or group
Purple	O or F	Flagella	Pol or Per	*Chromobacterium*
No pigment or green, yellow, red or orange	F	Flagella	Per	Enterobacteria
	F	Flagella	Pol	*Aeromonas*
	O or –	Flagella	Pol	*Pseudomonas*
	O or –	Flagella	Per	*Alcaligenes*
	O	Flagella	Shoulder	*Agrobacterium*
	O or F	Gliding	–	Myxobacteria
	O or –	–	–	*Acinetobacter*

O, oxidative; F, fermentative and oxidative; –, no action; Pol, polar, and Per, peritrichous.

A more detailed consideration of the identification of the organisms and also the relevant literature prior to 1966 was published by Park and Holding (1966). Additional comments relating to some of the more important publications since 1966 and any reconsideration of the taxonomic status of the genera and groups listed in Table 1 is appended below.

Alcaligenes *spp*

A revised description of the genus allows for the inclusion of bacteria able to oxidize glucose (Holding and Shewan, in press). Certain species previously allocated to the genus *Achromobacter* may now be included and there appears to be little justification for the retention of *Achromobacter*. Taxonomic aspects of *Alcaligenes* have also been discussed recently by Rhodes (1970).

Agrobacterium *spp*

Keane, Kerr and New (1970) and Allen and Holding (in press) have proposed alternative classifications of species. Except for plant pathogenic characteristics and the frequent occurrence of flagella in the shoulder (or sub-polar) position, differences between these organisms and *Alcaligenes* spp appear indistinct.

Myxobacteria

The taxonomy of the simple myxobacteria which produce only vegetative cells and no resting stage has been discussed recently by Lewin (1969). The types likely to be numerous in soil have been allocated to the genera *Cytophaga* and *Flexibacter*.

Acinetobacter *spp*

This distinct group of non-motile bacteria has been considered in detail by Thornley (1967) and Baumann, Doudoroff and Stanier (1968*a*, *b*). The latter authors allocate some members of the group to the genus *Moraxella*.

Whilst the characteristics discussed will permit the preliminary identification of many of the Gram negative heterotrophic bacteria it should be stressed that other groups, for example, *Bacillus* and *Arthrobacter* may contain Gram negative forms. Other groups should not therefore be overlooked in any comprehensive investigation of the soil Gram negative bacterial population.

References

ALLEN, O. N. & HOLDING, A. J. (in press). Genus *Agrobacterium*. Bergey's Manual of Determinative Bacteriology, 8th Ed. Baltimore, Md.: Williams and Wilkins Co.

BAUMANN, P. (1968). Isolation of *Acinetobacter* from soil and water. *J. Bact.*, **96,** 39.

BAUMANN, P., DOUDOROFF, M. & STANIER, R. Y. (1968a). Study of the *Moraxella* group. I. Genus *Moraxella* and the *Neisseria catarrhalis* group. *J. Bact.*, **95,** 58.

BAUMANN, P., DOUDOROFF, M. & STANIER, R. Y. (1968b). A study of the *Moraxella* group. II. Oxidase-negative species (genus *Acinetobacter*). *J. Bact.*, **95,** 1520.

CLARK, A. G. (1969). A selective medium for the isolation of *Agrobacterium* species. *J. appl. Bact.*, **32,** 348.

GRAHAM, P. H. (1969). Selective medium for growth of *Rhizobium*. *Appl. Microbiol.*, **17,** 769.

HOLDING, A. J. (1960). The properties and classification of the predominant Gram-negative bacteria occurring in soil. *J. appl. Bact.*, **23,** 515.

HOLDING, A. J. & SHEWAN, J. M. (in press). Genus *Alcaligenes*. Bergey's Manual of Determinative Bacteriology, 8th Ed. Baltimore, Md.: Williams and Wilkins Co.

KEANE, P. J., KERR, A. & NEW, P. B. (1970). Crown gall of stone fruit. II. Identification and nomenclature of *Agrobacterium* isolates. *Aust. J. biol. Sci.*, **23,** 585.

LEWIN, R. A. (1969). A classification of flexibacteria. *J. gen. Microbiol.*, **58,** 189.

NEW, P. B. & KERR, A. (1971). A selective medium for *Agrobacterium radiobacter* Biotype 2. *J. appl. Bact.*, **34,** 233.

PARK, R. W. A. & HOLDING, A. J. (1966). Identification of some common Gram-negative bacteria. *Lab. Pract.*, **15,** 1124.

RHODES, M. E. (1970). Aniline utilization by *Alcaligenes faecalis*: a taxonomic reappraisal. *J. appl. Bact.*, **33,** 714.

SANDS, D. C. & ROVIRA, A. D. (1970). Isolation of fluorescent pseudomonads with a selective medium. *Appl. Microbiol.*, **20,** 513.

SCHROTH, M. N., THOMPSON, J. P. & HILDEBRAND, D. C. (1965). Isolation of *Agrobacterium tumefaciens—A. radiobacter* group from soil. *Phytopathology*, **55,** 645.

THORNLEY, M. J. (1967). A taxonomic study of *Acinetobacter* and related genera. *J. gen. Microbiol.*, **49,** 211.

Vineyard Yeasts—an Environmental Study

R. R. Davenport

University of Bristol, Department of Agriculture and Horticulture, Research Station, Long Ashton, Bristol, England

A detailed study of the vineyard, established at the Research Station in 1965, has shown the presence of several groups of microorganisms including yeasts and yeast-like forms. Among this microflora were many potential spoilage organisms such as *Schizosaccharomyces* spp and acetic acid bacteria, but fermenting yeasts (*Saccharomyces* spp) were rarely found. This is contrary to results of studies made on most other European vineyards where the latter yeasts are reported as predominant.

The results of this microbial ecological survey were only possible because of the formation of a model system in which the vineyard was considered in relation to the surrounding biosphere and the timing of the microbiological sampling was related to the development stages of the grape vine. Further, sources of yeasts within the vineyard and its perimeter were determined with reference to microclimate, invertebrate vectors and vineyard husbandry.

Objective

The objective of this study was a systematic, quantitative survey of the yeasts and yeast-like microorganisms present in the vineyard at Long Ashton. Thus, investigations were undertaken to see whether the microflora was similar or not to that of any foreign vineyard.

(*a*) Whether it was unique or merely a reflection of the microflora of the surrounding orchards, trees and a willow hedge.

(*b*) The effectiveness of various vectors in the dissemination of yeasts and yeast-like species between them. To assess these possibilities the vineyard was considered as an ecosystem, forming part of the biosphere and thence divided into three zones (Davenport, 1970*a*): (1), atmosphere—to include yeasts borne on the wind, wind-borne seed heads and flying vectors; (2), phyllosphere—comprising all aerial vine parts and zones immediately around those parts, and (3), rhizosphere—embracing root surfaces of the vine and of other vegetation, the soil and its macrofauna. In this section the rhizosphere will be considered in detail with only brief reference to the atmosphere and phyllosphere.

The Site

Geology

The geology of the vineyard area is complex. It is composed of brown earth made up of a mixture of three soil series, namely, Lulsgate, Swindon bank and Wrington, forming a shallow layer on hard rock, composed of carboniferous limestone, millstone grit and dolomitic conglomerate (Cope, 1969). When the vineyard was formed the whole area was levelled, thus disturbing the earth and a mediaeval wall system (S. H. Legge, pers. comm.) just beneath the surface.

Geography

The vineyard at Long Ashton Research Station (latitude, 51°25′N; longitude, 2°40′W) is situated on a south facing slope bordering a shallow valley which runs approximately east to west. The altitudes of the northern and southern slopes are 450 and 250 ft respectively. Figure 1 shows the location of the grape vines and their surroundings, the whole area given the general term, vineyard.

Microclimate

Detailed studies of temperature and wind were made within and between the vines, coupled with climatic observations at three standard agrometeorological stations (Anon, 1969), sited as shown on Fig. 1.

The warmest area in the vineyard occurred approximately in the centre, while the coolest area was on the western edge, a little north of the centre row near the adjoining willows. The greatest deviations in temperature recordings occurred when the weather was calm and clear (Davenport, 1970a). For example, extremes of temperature measured at ground level and 3 ft up in the vines were 15·5° and 7° respectively. It was also shown that a shelter belt was formed by the elms and willows on the WSW perimeter, so that the maximum air temperatures in the top and bottom of the vineyard could vary by 4–5°. According to Winkler (1965) temperature is the most important climatic factor for grape growing. He also stated that the vine remains dormant until the mean daily temperature reaches c. 50°F (10°). This figure was used as a base line for calculating heat summation in different areas of the vineyard.

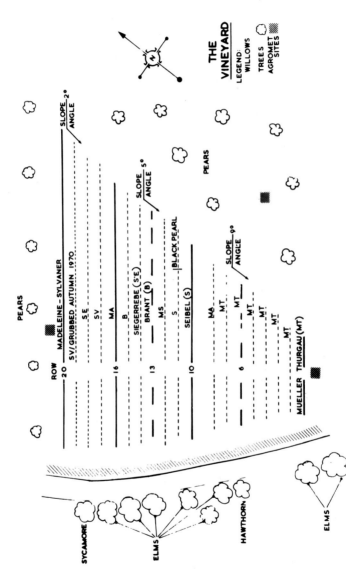

FIG. 1. Map of the vineyard showing the adjacent orchards, trees and willow hedge all considered as part of the ecosystem.

Heat summation

Heat summation is calculated for the period, bud break to the date of harvesting the fruit (Winkler, 1965). Thus:

$$\frac{\text{Maximum} + \text{Minimum} \ (^\circ\text{F})}{2} -50^\circ\text{F} \times \text{period (in days)} = \text{Degree days}$$

e.g. for a 30-day period when the maximum and minimum temperatures were 80 and 60°F respectively, the heat summation would be:

$$\frac{80 + 60}{2} = 70 \text{ Mean temperature}$$

$$70-50 = 20$$
$$20 \times 30 = 600 \text{ Degree days for 30-day period}$$

In this project, values for Degree days were used to compare heat summation distribution between seasons, vertical and horizontal levels in the vines and the agrometeorological stations.

Sampling and Processing

A brief note about the terms monitoring and sampling is made because it could be inferred that the one term is synonymous with the other, whereas in this exercise there are distinct meanings defined for each. Monitoring means measurement of the physical, chemical and biotic factors within the vineyard habitat; sampling relates to the collection of samples which were later processed for yeasts and yeast-like organisms. In some cases, e.g. in microfauna studies, the same sample was used both as a sample and part of the monitoring programme.

Monitoring

Simultaneously with the systematic collection of samples at each stage of grape vine development, vineyard husbandry practices were noted and the position of weeds, bird perch sites and animal vectors were mapped. Soil analyses were also carried out together with observations obtained from the 3 agrometeorological stations (Fig. 1) and within the vine growing area.

Sampling

The stages of grape vine development occurred in the following months at Long Ashton Research Station:

Dormant bud	November–April
Early shoots	April–July

Flowering–fruit set   July–August
Immature fruit ⎱ I   August–September
               ⎰ II   September
Mature fruit   September–October

Thus, because a biological rather than a lunar calendar was used, direct comparisons could be made between the yeast microfloras at similar stages of grape vine development, irrespective of differences due to season or location (Davenport, 1968).

Two types of sample were taken, first a preliminary bulk sample and later, a detailed sample from a more restricted area. The bulk sample, taken at random from at least 20 points, was mixed and examined microbiologically and chemically. Analyses of these bulk samples gave a broad indication of the numbers and types of microorganisms present and any chemical data which may influence the microorganisms present. Hence the programme of detailed samples was prepared from a consideration of the results obtained from the bulk samples and the nature of the habitat. The soil and soil vector studies given here serve to illustrate part of the integrated study of the micro-ecology as applied to an English vineyard.

Soil

Soil samples were taken from the points shown on Fig. 2, by an auger previously sterilized by dipping the screw-end into alcohol and flaming. Soil was only taken from the top 5–10 cm layer, where most soil yeasts reside (Phaff, Miller and Mrak, 1966; Davenport, 1968). Four cores from each sampling point were placed in a pre-weighed sterile tin, sealed and processed as soon as possible.

Soil macrofauna

Two methods were employed for sampling soil macrofauna:

A metal frame quadrat (Raw, 1959). Areas of 0·5 m² were searched to a depth of 10 cm, using a flamed garden hand fork and forceps.

Pitfall traps. These traps consisted of a glass jam jar (Oldroyd, 1958) and an aluminium rain shelter top which were sterilized by autoclaving (121°/15 min) and then placed in a permanent site made from a length of plastic drainpipe. This was buried in the soil so that ground crawling vectors fell into the jars but were unable to climb the smooth surface of the glass (Fig. 3). Pitfall traps were set out in the vineyard and left for varying periods, depending on the season and experiments being carried out. Traps

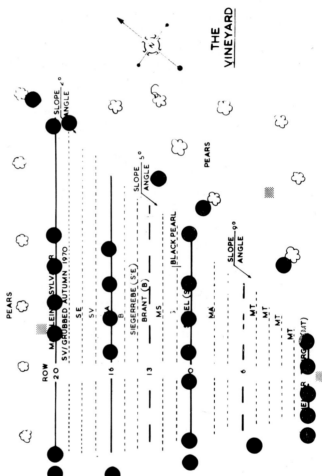

Fig. 2. Soil sampling points.

b

a

FIG. 3. (a) Pitfall trap, and (b) pitfall trap *in situ*.

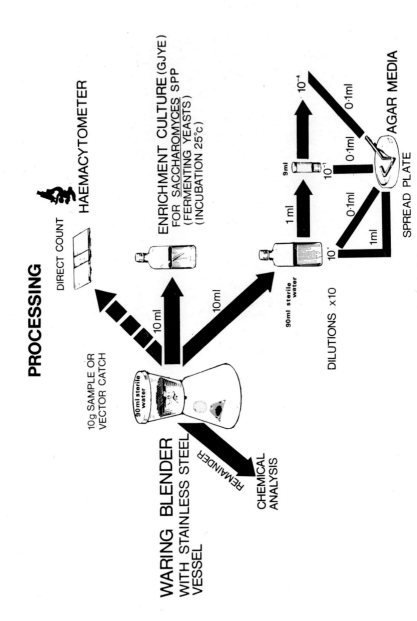

Fig. 4. Processing schedule.

were collected more often in warmer weather because of the observed cannibalism between species, as well as the possibility of digestion of yeasts within the alimentary tract of the vector (Phaff *et al.*, 1966). The vector catches were identified only or identified and subsequently processed to give an indication of the vector and the yeast population dynamics. Some vector specimens were allowed to crawl over the surface of nutrient agar in Petri dishes, while other vectors were dissected aseptically and part of the gut contents examined microscopically (Phaff *et al.*, 1966). The remainder of the alimentary canal was placed in Grape Juice Extract medium (Davenport, 1970*a*) or streaked across a plate of solid medium (Beech and Davenport, 1971). After *c.* 1 week incubation, the enrichment culture was examined microscopically and plated out.

Processing

Samples were macerated and thoroughly mixed using the Waring blender method described by Beech and Davenport (1971). To prevent yeast growth during comminution, chilled sterile water was used for the maceration process. Spread plates of tenfold dilutions of the macerate were incubated until the yeast colonies are well defined (Beech and Davenport, 1971). The complete scheme for processing, media used and incubation conditions is given in Fig. 4 and Tables 1 and 2.

TABLE 1. Media used for isolating vineyard microorganisms

Media*	pH used	Comments
Grape Juice Yeast Extract	3·5 and 4·8	A general purpose rich nutrient medium; good morphological differentiation for most yeasts; all bacteria suppressed except acid tolerant groups
Potato dextrose (Oxoid)	3·5 and 5·6	Good growth of mucoid colonies of *Cryptococcus* and certain carotenoid strains; *Lipomyces* species isolated only from this medium
Littman ox-gall (Oxoid)	7·0	Restricts spreading of fungal colonies and inhibitory to some bacteria; many *Candida* spp grow well on this medium

Media*	pH used	Comments
MRS (Oxoid)	6·2	Fungal colonies very much reduced in size; lactobacilli and other bacteria grow well; fission and certain carotenoid yeasts only isolated from this medium
Nutrient agar (Oxoid) containing 0·5 % (w/v) glucose	7·4	Bacteria and pH sensitive yeasts

* Formulae given in the Appendix.

TABLE 2. Incubation temperatures used

Temperature (°C)	Time	
2	1–3 months	
5	1–3 ⎫	
15	1–2 ⎬ weeks	
25	1·2 weeks	enrichment cultures for fermenting species only
37*	5 days	10°/1 ml only

* Screening for possible pathogenic and/or thermophilic strains.

Identification of Isolates

Methods

Isolation plates were examined quantitatively and qualitatively for all microorganisms present. The macroscopic and microscopic appearance of at least one of each different colony form was described so that the microorganisms could be put into groups of either bacteria, fungi, yeasts and/or yeast-like organisms. The latter 2 groups were treated in greater detail and subdivided into smaller divisions based on a modified key devised by Beech, Davenport, Goswell and Burnett (1968). Similar isolates, together with selected type species, could then be identified using the methods outlined in Tables 3 and 4, and selected tests from Lodder (1970), Barnett (1971) and Campbell (1971).

TABLE 3. Tests used in yeast identification

Characteristics	Determinative methods/notes	References
Pigment, macro- and micro-morphology: esters	Isolation and purification media: malt wort agar and dalmau plates (cornmeal agar)	Walters (1943) Wickerham (1951)
Ascospores and ballistospores	During isolation and purification procedures	Davenport (1968)
Cell sizes, pellicle and ester formation	Malt wort liquid	Kreger-van Rij (1964)
Fermentation	Durham tube method	Wickerham (1951)
Assimilation (carbon compounds)	Liquid Solid/modification of method	Wickerham (1951) Beech (1957)
Assimilation	Liquid Solid/modification of liquid method	Wickerham (1951)
Osmotic pressure	Moderate, liquid and solid 50% (w/w) glucose	Wickerham (1951) Kreger-van Rij (1964)
Growth at 37°	Malt agar Liquid, glucose assimilation test	Kreger-van Rij (1964) Wickerham (1951)
Acid production	Solid	Custers (1940)
Gelatin liquefaction	Nutrient gelatin medium (Difco)	
Starch production	Final carbon assimilation tubes tested with Lugol's iodine	Wickerham (1951)
Vitamin deficiency test	Liquid and solid	Wickerham (1951)

TABLE 4. Potential industrial spoilage strains. Supplementary identification tests for yeasts and yeast-like organisms

Test/media	Authors
Growth in ethanol* 8, 12 and 16% (v/v) in Grape Juice Yeast Extract Growth in grape juice at pH 3·5	Campbell (1971)
Lypolysis/tributyrin agar (Oxoid)	Davenport (1968) Campbell (1971)
Actidione (cyclohexamide) resistance	Lodder (1970)
Assimilation of malic acid	Beech (1957) Davenport (1968, 1970a)
Assimilation of chlorogenic acid	Davenport (1970a)

* Campbell (1971) used malt extract broth not grape juice.

Distribution of microorganisms

Tables 5, 6, 7 and 8 give the groups of microorganisms, yeasts and yeast-like organisms isolated from the vineyard.

TABLE 5. Groups of microorganisms isolated from an English vineyard

Yeasts

Dark	Mucoid	Apiculate
Carotenoid	Smooth	Fission
Pulcherrimin	Rough	

Fungi

Alternaria	*Cladosporium*
Aspergillus	*Mucor*
Botrytis	*Pencillium*
Cephalosporium	Sterile forms

Bacteria

Acetic acid	Pigmented
Lactic acid	Non-pigmented

TABLE 6. Distribution of yeasts and yeast-like organisms within an English vineyard

Organisms	Atmos-phere	Phyllos-phere	Rhizos-phere
Dark			
Aureobasidium pullulans	+	+	+
Cladosporium*	+	+	+
Cryptococcus	−	+	−
Oosporidium	−	−	+
Trichosporon†	−	−	+
Trichosporonoides	−	+	−
Carotenoid			
Cryptococcus	+	+	+
Rhodotorula†	+	+	+
Rhodosporidium	+	+	+
Sporobolomyces	+	+	+
Sporidiobolus	−	+	−
Pulcherrimin			
Metschnikowia	−	+	+
Mucoid			
Bullera	−	+	+
Cryptococcus†	+	+	+
Lipomyces	−	−	+
Leucosporidium	−	+	−
Hansenula†	−	−	+

Organisms	Atmos-phere	Phyllos-phere	Rhizos-phere
Smooth and/or Rough			
Candida†	+	+	+
Debaryomyces†	+	+	+
Endomycopsis	−	+	+
Hansenula†	+	+	+
Metschnikowia	+	+	+
Saccharomyces†	−	−	+
Torulopsis†	+	+	+
Apiculate			
Hanseniaspora†	−	+	+
Kloeckera†	−	+	+
Nadsonia†	−	+	+
Fission			
Schizosaccharomyces†	−	+	+

* Yeast-like form.
† Positive to one or more of the screening tests for industrial spoilage organisms.

TABLE 7. Examples of pigmented yeasts and yeast-like species isolated from an English vineyard

Organisms	Distinguishing features for species within the given group		
Dark			
Aureobasidium pullulans	Multilateral budding; some irregular cells but no arthrospores; septa formed sometimes with old cells		
Oosporidium margaritiferum	Multilateral budding on a broad base; septa formation but no arthrospores		
Trichosporon pullulans	Multilateral budding; true mycelium and arthrospores; lipase activity		
Carotenoid	Ballistospores	Mycelium	Nitrate assimilation
Rhodotorula rubra	−	−	−
Rh. glutinis	−	−	+
Sporobolomyces roseus	+	−	+
Sp. pararoseus	+	−	−
Sp. salmonicolor	+	+	+
Pulcherrimin			
Metschnikowia pulcherrima	Glucose fermentation only; characteristic asci produced		

TABLE 8. Examples of non-pigmented yeasts and yeast-like organisms isolated from an English vineyard

Organisms	Distinguishing features for species within the given group
Mucoid	
Cryptococcus albidus	Ascospores not formed
Lipomyces starkeyii	Ascospores formed
Rough	
Candida humicola	Abundant true and pseudomycelium characteristic branching for this species
C. parapsilosis	Fermentation of glucose and galactose; wrinkled colonies
C. valida	Weak fermentation of glucose; powdery colonies
Metschnikowia reukaufii	Glucose not fermented; no pulcherrimin pigment; characteristic asci produced
Smooth	
Debaryomyces hansenii	Ascospores formed
Torulopsis inconspicua	Ascospores not formed
Apiculate	
Hanseniaspora valbyensis	Hat-shaped ascospores
Han. uvarum	Helmet-shaped ascospores
Kloeckera apiculata	No ascospores; maltose not assimilated
Nadsonia elongata	Characteristic mother-daughter cell ascospore formation
Nadsonia sp. *A*	As *N. elongata*, also nitrate assimilated
Kloeckera corticis	No ascospores; maltose assimilated
Fission	
Schizosaccharomyces sp. *X*	No fermentation

Discussion

Alexander von Humboldt, writing in 1807 on plant geography (Scamoni, 1966), said: "In the great chain of causes and effects no thing, and no activity should be regarded in isolation." This should be the fundamental basis of modern ecological concepts. Sometimes, however, ecological studies are merely a description of a given habitat or area and a catalogue of unrelated features. This catalogue is not sufficiently precise for an integrated study, especially when one considers that a habitat is not an isolated point in space but a selected area which is part of the planet earth. The boundary of this area is rarely definite, usually one habitat merges unevenly into one or more neighbouring habitats. Hence one must first map the area and note any possible boundary interactions as well as define the precise meaning of any terms used in its description and subsequent examination.

    This chapter attempts to describe an exercise in microbial ecology which forms part of a "model system" used for examining a section of a given environment, in this case an English vineyard ecosystem. Tansley's (1935) definition of the term ecosystem is designated as a system resulting from the integration of all living and non-living factors of the environment. In this context the term implies a concept and not a unit of landscape, that is, one should look beyond the particular entity to be studied and must consider the interrelationships among the components of the landscape and their environment (Van Dyne, 1969). However, one must contain the application of ecological studies within manageable proportions; this means formulating a concise objective and mapping and monitoring according to the facilities available. As the project advances then there must be a continuous assessment of the data so that one can either extend or retract certain avenues of work, in order to reach efficiently the desired objective.

Many methods were used to examine the zones of the vineyard, so that it is now possible to state, in general terms, the distribution of the dominant yeasts and yeast-like organisms, whose physiology determines whether they are part of the resident or transient microfloras (Davenport, 1968) within the vineyard ecosystem (Davenport, 1972).

	Atmosphere	**Phyllosphere**	**Rhizosphere**
Quantitative ↑	Carotenoid genera	*Aureobasidium* strains	*Trichosporon* spp
		Carotenoid genera	*Aureobasidium* strains
			Carotenoid genera

Dominance —————————————————————————————→

Qualitative

However, there were transition areas between the zones where the microflora was differentiated less clearly. Vine bark samples included a mixture of soil and leaf microorganisms such as *Aureobasidium* and *Trichosporon* strains as well as an unidentified "black yeast" not found elsewhere.

Microclimatological observations were not only of value for determining sites of possible ecological interest, but also were of practical value for vineyard husbandry. Fruiting canes of the grape vines can be grown equally well at 1, 2 or 4 ft levels above the ground since there was no difference in the Degree days (average value, 1760 Degree days) at the heights when measured over the whole season. There were some differences in heat unit distribution according to the row (Fig. 5) and individual points within the row which could be correlated with possibly uneven grape berry ripening and concentration of macrofauna. At present the seasonal data show the extreme points of climatic variation which enables one to monitor and sample the most likely areas of possible ecological significance.

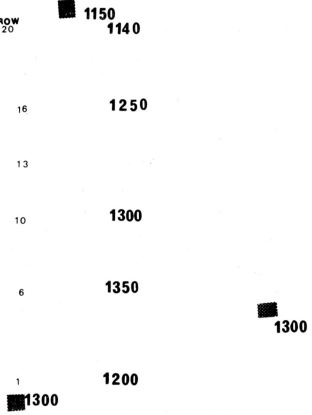

FIG. 5. Heat units within the vineyard. Average values for the whole row at 1, 2 and 4 ft levels, from dormant bud to mature fruit (1970) are given.

The soil map (Fig. 2) indicates the large numbers of sampling points required because of the heterogeneous nature of the soil in this area; had the location been all of one soil type then a smaller number of sampling points, chosen at random, would have been sufficient. Extensive monitoring showed that in general the soil in the lower half of the vineyard was neutral to alkaline, becoming progressively more acid moving up the slope to the northern boundary (Davenport, 1970b). The pH profile of the vineyard is illustrated by Fig. 6, which also gives the areas of shallow and deep acid and alkaline soil.

Having obtained this information, the numbers of soil samples were reduced to give a comparison between shallow and deep acid and alkaline soils. Monitoring of the other sampling points was continued so that any

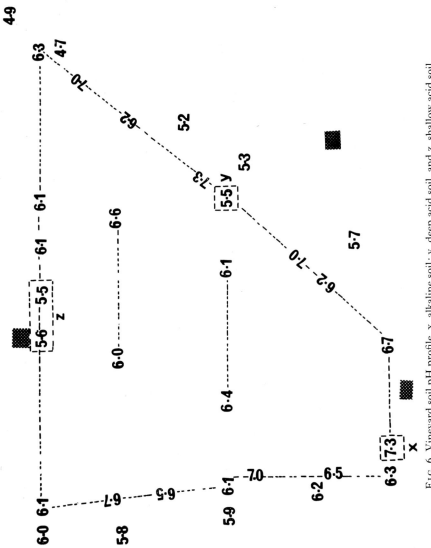

FIG. 6. Vineyard soil pH profile. x, alkaline soil; y, deep acid soil, and z, shallow acid soil.

change in pH could be noted and linked with an immediate microbiological investigation.

There were very few vine rootlets growing in acid soil but they were numerous in alkaline soil, therefore rootlets growing in both types of soil were examined microbiologically. The microflora of both contained bacteria and fungi, including species of *Aspergillus*, *Cephalosporium*, *Cladosporium*, *Mucor*, *Penicillium* and unidentified forms, and also yeasts belonging to the genera *Trichosporon* and *Candida* (Davenport, 1972), whereas species of *Bullera*, *Lipomyces*, *Saccharomyces*, *Schizosaccharomyces* and an unknown apiculate yeast were only found in the acid zones. In Table 9

TABLE 9. Numbers of yeasts from soils, rootlets and soil invertebrate vectors

	Range of counts ($\times 10^3$)	
	Acid	Alkaline
Soils	4–300	20–50
Rootlets	1–10	0·2–1·0
	Surfaces	
	External	Internal
Soil invertebrate vectors	2·2–TM*	1·6–TM

* TM, too many to count.

it can be seen that the counts of microorganisms from the acid and alkaline soils and rootlets were different, including the isolation of a *Saccharomyces* sp, which was found on only one occasion (< 1 cell/g soil) from the shallow acid soil shown on Fig. 6.

It was also observed that the type of weeds growing in the vineyard area could be correlated either with soil moisture, pH or distribution of ground microfauna (Fig. 7). Weeds were important because:

(a) Weed flowers and seed-heads contained a variety of possible industrial spoilage microorganisms: those on the flowers depended on flying vectors for their dissemination and those on the seed-heads on the wind. Thus plants of different physiological age to the vines and some considerable distance away can have an effect on the vine microflora.

(b) It has been shown recently that excretions from the roots of certain ubiquitous weeds, especially milk thistle (*Sonchus arvensis*), can affect the grape vine root system (Racz and Slaba, 1971) which, as with other physical and/or chemical factors such as pH, may also influence the soil microflora. It follows that different organs of the weeds, together with climatic factors and vectors, contribute to the transient as well as the resident microflora of the vineyard.

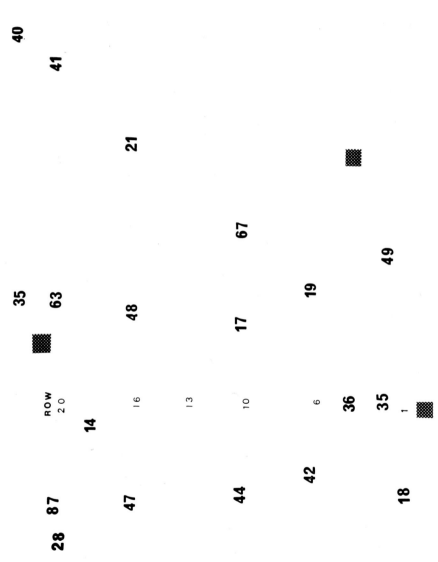

Fig. 7. Distribution of ground vectors from pitfall traps in the vineyard during the period, immature fruit to dormant bud, 1970. NB: The largest number of vectors occurred in the areas of greatest moisture and/or weed cover.

Results obtained with counts both on the macrofauna and microflora, proved conclusively that yeast populations were carried both externally and internally (Fig. 8). Viable yeast-like organisms were found even inside the lumen of stinging nettle (*Uritica dioica*) (Fig. 9), previously ingested by a grasshopper (Davenport, 1972). A carabid beetle (*Carabus violaceus*) was responsible for carrying yeasts and yeast-like organisms, including a *Nadsonia* sp, isolated for the first time in an English environment, from the willow hedge to the adjacent edge of the vineyard (Davenport, 1972).

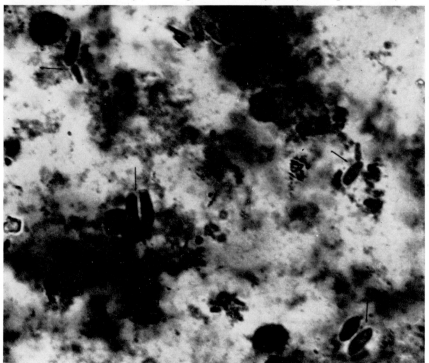

FIG. 8. Budding yeast cells in beetle gut contents. × 1000.

A further association was confirmed between the rare, ascomycetous yeast, *Metschnikowia* (Fig. 10) and a flying vector identified tentatively as *Proctrupidae* spp. This was confined to the NE vineyard perimeter (Davenport, 1972) where certain beetle larvae and adults were located. As members of the *Proctrupidae* can parasitize other insects (Borradaile, Potts, Eastham and Saunders, 1959), it could be that there is an association between vectors, yeasts and habitat in this instance.

Similarly vineyard husbandry had an influence; a *Saccharomyces* sp was isolated from mummified palmette pears (Fig. 1) found on the ground some months previous to the isolate mentioned earlier. It could be specu-

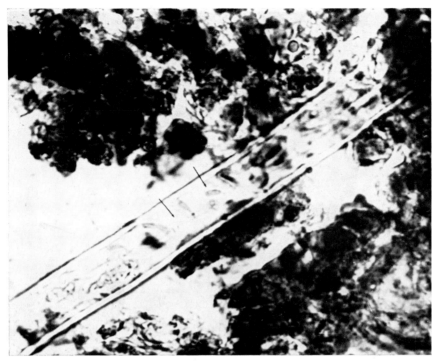

FIG. 9. Frass contents. × 1000.

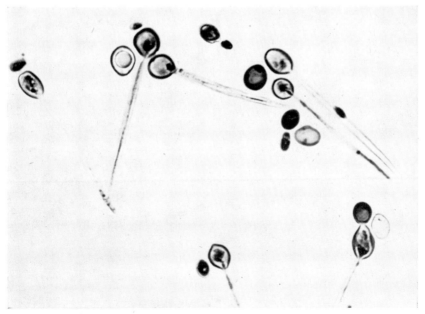

FIG. 10. *Metschnikowia* sp. × 1000.

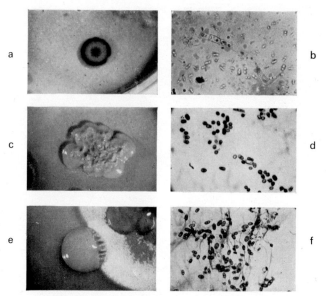

FIG. 11. (a) *Aureobasidium pullulans* colony. (b) *A. pullulans* cells. Note endospores in mycelium. (c) *Trichosporon* sp colony. (d) *Trichosporon* sp cells. (e) *Rhodotorula glutinis* colony. (f) *Rhodotorula glutinis* cells. Note formation of extra-cellular polysaccharide and intra-cellular lipids. (a), (c), (e) ×2·5; (b), (d), (f) ×600.

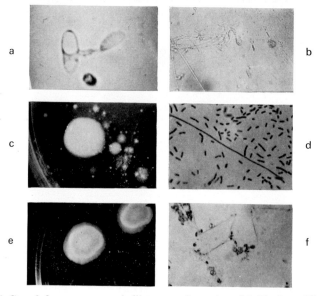

FIG. 12. (a) *Sporobolomyces roseus*—ballistospore formation. (b) *Rhodosporidium* sp. Note teliospores, promycelium and clamp connections. (c) *Lipomyces starkeyi* colony. (d) *Schizosaccharomyces* sp cells. (e) *Metschnikowia* sp. (f) *Metschnikowia* sp. Needle ascospores, green, and vegetative cells, red. (a), (b), (d), (f) ×600; (c), (e) ×2·5.

lated that vectors were responsible for transferring this yeast from the mummified fruit, which have been shown to harbour large populations of yeast and acetic acid bacteria (Davenport, 1968). Mummified grapes left on the vines until the next flowering season were also found to contain spoilage microorganisms such as apiculate and film yeasts as well as acetic acid bacteria. As these microorganisms were found not to survive the winter months other than in mummified fruit, it was concluded that the high sugar content of the fruit made this a special niche for their survival. It was observed also that leafy summer prunings left on the vineyard floor developed very heavy growth of fungi, particularly *Botrytis cinerea*, which can cause heavy infection on the grapes. Ideally therefore one should remove mummified fruits and prunings from a vineyard to prevent the propagation of vine diseases and wine spoilage organisms.

The objective of this study is almost complete and will be fully discussed elsewhere but present results indicate that the vineyard has acquired a similar flora to that of the surrounding orchards, trees and willow hedges, together with some other microorganisms found only on the vines. If in the future the latter survive on the vines and not in the neighbouring areas, then one could conclude that the vineyard has acquired a unique microflora.

Davenport (1970*a*) summarized this project by stating that "the vineyard microflora sampled at any biological time is dependent on the physiological age of the plants in its immediate vicinity, on the micro-climate which enables the microflora to be distributed by ground and air vectors, and on the vineyard topography". It was further stated that "the survival of the components of the microflora is determined by their combined physiology and the interaction of the individual constituents and environmental conditions".

Acknowledgements

The author is indebted to Mr E. G. R. Chenoweth for preparing Fig. 6, and Mrs J. Sadd for assistance with processing the photographic plates.

*Supplements to media**
Generally all dilutions were plated in triplicate, including one set of dilution plates of Grape Juice Yeast Extract and Potato Dextrose media with the addition of anti-bacterial and anti-fungal agents. Iron (0·5% (w/v) ferric ammonium citrate) and biotin was also added so that red-maroon colonies of the pulcherrimin producing yeasts could be distinguished from pink-orange colonies of the carotenoid yeasts by the solubility of pulcherrimin in methanolic potassium hydroxide but not in organic solvents (van der Walt, 1952).

* Additional details and methods are discussed by Beech and Davenport (1969, 1971).

G

Appendix

Media used for the isolation of vineyard yeasts

Media	Composition*	Sterilization (temperature/min)	Reference
Grape juice yeast extract (GJYE)	Yeast extract, 1·0; thiamine, 0·1 mg, and agar† 3·0. pH 4·8	115°/15	Davenport (1970a)
Potato dextrose—CM139—(PD)	Potato extract (Oxoid L 101), 0·4; dextrose, 2·0, and agar, 1·5. pH 5·6	121°/15	Oxoid (1965)
MRS (CM359)	Peptone, 1·0; Lab Lemco, 0·8; yeast extract, 4·0; dextrose, 2·0; "Tween" 80, 0·1 ml$+§$, and agar, 2·0. pH 6·2	121°/15	Oxoid (1969)
Storage medium (MYPG)	Malt extract, 0·3‡; yeast extract, 0·3; Bacto peptone, 0·5; glucose, 1·0, and agar 2. pH 6·8	121°/15	Wickerham (1951)
Nutrient/glucose (NAG)	Glucose, 0·25; yeast extract, 0·2; Lab Lemco (Oxoid), 0·1; Bacto peptone, 0·5; sodium chloride, 0·5, and agar 2·0. pH 7·0	115°/15	Oxoid (1965)
Littman—CM91—(L)	Peptone, 1·0; dextrose, 1; Ox-gall, 1·5; crystal violet, 0·001, and agar, 1·5. pH 7·0	121°/15	Oxoid (1965)

* Amounts, g/100 ml distilled water; pH adjusted with KOH pellets or $N-HCl$ and checked with pH meter (ELL Model 23A).
† Ionagar No. 2 (Oxoid).
‡ Spray malt "A" (Munton and Fison, Suffolk).
§+ di-potassium hydrogen phosphate, 0·2; sodium acetate $3H_2O$, 0·5; triammonium citrate, 0·2; magnesium sulphate $7H_2O$. 0·2, and manganese sulphate $4H_2O$, 0·005.

Media used in yeast identification tests

Media	g/100 ml	Sterilization	Reference
Washed agar	Ionagar No. 2 (Oxoid), 4. pH not adjusted	120°/15 min	Beech (1957)
Cornmeal	Cornmeal (Difco), 1·7. pH not adjusted		Wickerham (1951)
Arbutin agar	Arbutin, 0·5; yeast extract, 1·0, and agar, 2·0. pH not adjusted	115°/15 min	Lodder and Kreger-van Rij (1952)
Acetate agar	Acetate, 0·4, and agar 1·5. pH 6·5		Fowell (1952)
Chalk agar	Yeast extract, 1·0; glucose, 5; calcium carbonate, 0·5, and agar, 2·0. pH not adjusted	115°/15 min	Custers (1940)
Keyln +	Yeast extract, 1·0; Bacto tryptone, 0·25; sodium chloride, 0·062, and agar 2·0. pH not adjusted	115°/15 min	Kleyn (1954)
Malt wort	Spray dried malt, 20, with or without agar (4·0). pH 5·4	Steamed–3 d 115°/15 min	Walters (1943)
Moderate osmotic pressure	Glucose, 5; sodium chloride, 10, and agar, 3·0	Filter sterilized 115°/15 min	Wickerham (1951) Davenport (1968)
50% Glucose	Glucose, 50; yeast extract, 1·0, and agar, 3·0. pH not adjusted	115°/15 min	Kreger-van Rij (1964)
Nutrient gelatin	Bacto beef extract (Difco), 0·3; Bacto gelatin, 12, and Bacto peptone 0·5	115°/15 min	
Vitamin deficiency test	Yeast vitamin free base,* 16·7. pH not adjusted	Filter sterilized	Wickerham (1951)
Nitrate	Yeast carbon base, 0·078, and potassium nitrate.* pH not adjusted	Filter sterilized	Wickerham (1951)
Carbon assimilation tests	Yeast nitrogen base + carbon sources.* pH 6·8	Filter sterilized	Wickerham (1951)
Fermentation	Yeast extract, 0·45, peptone, 0·75 + bromothymol blue	115°/15 min	Wickerham (1951)

* Tetrazolium salt (2:3:5—triphenyl—tetrazolium chloride; TTC) BDH. Modified method using washed or noble agar (Difco) + TTC

*Simplified morphological and physiological key of multipolar budding yeasts and yeast-like genera**

Macro-morphological group / Organisms	Micro-morphology (cells)					Physiology		Comments
	Myce-lium	Arthro-spores	Circu-lar	Oval to cylin-drical	Mix-ture	Fer-men-tation	Nitrate assimi-lation	
Dark								
Aureobasidium	+	−	+	+	+	−	±	Many colonial forms
Cladosporium	−	−	+	+	+	NT	NT	Yeast-like phase of fungus
Oosporidium	+	−	+	+	+	−	+	Chain formation and septa independent of budding
Trichosporon	+	+	−	+	+	±	±	Budding cells of various shapes
Trichosporonoides	+	+	+	+	+	+	+	Conidia produced
Carotenoid								
Cryptococcus	−	−	+	−	−	−	±	Starch-like compounds synthesized; inositol assimilated
Rhodosporidium	+	−	−	+	+	−	+	Clamp connections and teliospores
Rhodotorula	−	−	+	+	−	−	± P	Yeast cells only
Sporidiobolus	±	−	−	+	−	−	+ P	Clamp connections and ballistospores
Sporobolomyces	±	−	+	+	+	−	± P	Ballistospores; no clamp connections
Non-carotenoid								
Hansenula	+	−	+	+	+	±	+	Pink colonies—one or two species only
Oosporidium	+	−	+	+	+	−	+	Chain formation and septa independent of budding

Macro-morphological group / Organisms	Micro-morphology (cells)					Physiology		Comments
	Mycelium	Arthrospores	Circular	Oval to cylindrical	Mixture	Fermentation	Nitrate assimilation	
Pulcherrimin								
Candida	+	−	+	+	+	+	−	Imperfect state of *M. pulcherrima*
Kluyveromyces	±	−	+	+	+	+	−	External source of vitamins required
Metschnikowia	−	−	+	+	+	+	−	Characteristic needle-shaped ascospores
Mucoid								
Bullera	−	−	−	+	−	−	−	Ballistospores may be formed
Cryptococcus	−	−	+	+	−	−	±	Starch-like compounds synthesized; inositol assimilated
Hansenula	+	−	+	+	+	±	+	Esters produced; pellicle formation
Leucosporidium	+	−	+	+	+	±	+	Clamp connections not formed, growth usually below 15°
Lipomyces	−	−	+	+	−	−	±	Lipid and polysaccharide production
Pachysolen	−	−	+	+	−	+	+	Found only in tannin liquids
Torulopsis	−	−	+	+	−	±	+	No pseudomycelium
Smooth								
Citeromyces	−	−	+	−	−	+	+	Narrow base budding; heterothallic species
Cryptococcus	−	−	+	−	−	−	±	Starch-like compounds synthesized; inositol assimilated
Torulopsis	−	−	+	−	−	±	− NP	No pseudomycelium
Trigonopsis	−	−	−	−	−	−	− NP	Triangular shaped cells
Schwanniomyces	−	−	+	−	+	+	−	Egg shaped cells; all known isolates from the soil

Macro-morphological group / Organisms	Micro-morphology (cells)					Physiology		Comments
	Myce-lium	Arthro-spores	Circu-lar	Oval to cylin-drical	Mix-ture	Fer-men-tation	Nitrate assimi-lation	
Rough								
Endomycopsis	+	−	+	+	+	±	± NP	Abundant mycelium, arthrospores rare
Oosporidium	+	−	+	+	+	−	+ NP	Chain formation and septa independent of budding
Candida	±	−	+	+	++	±	±	Pseudomycelium always present
Debaryomyces	−	−	+	+	++	±	−	Pellicle formation, pseudomycelium sometimes formed
Rough and Smooth								
Hansenula	++	−	+	+	++	++	+	Esters produced; pellicle formation
Kluyveromyces	−	−	+	+	++	++	−	External source of vitamins required
Lodderomyces	−	−	+	+	++	±	−	Higher paraffins usually utilized
Metschnikowia	−	−	+	+	++	±	−	Characteristic ascospores
Nematospora	−	−	+	+	+	+	−	Septate branched hyphae; rarely found except on hard hazelnuts
Pichia	±	−	+	+	+	±	−	Pellicle and pseudomycelium formation
Saccharomyces	−	−	+	+	+	+	−	Fermentation of several sugars

* Based on data given by Beech *et al.* (1968). P, pigmented; NP, non-pigmented; and NT, not tested.

Anti-bacterial mixture

Actinomycin, 2 ppm, and aureomycin, 50 ppm. (Beech and Carr, 1960.)

Anti-fungal agents

Calcium propionate, 0·025% (w/v). (Lund, 1956; Bowen, 1962.)

Anti-yeast agent

Mixture added to facilitate the counting of acid tolerant bacteria.
Actidione, 10 ppm, and 8-hydroxyquinoline, 250 ppm.

Yeast genera with special cultural requirements

Coccidiascus	Not as yet grown in culture: observed only as a pathogen for certain insects
Pityrosporum *Saccharomycopsis* }	Special conditions required for culturing

References

ANON (1969). Met. 0.805 Observer's Handbook. London: HMSO.

BARNETT, J. A. (1971). Selection of tests for identifying yeasts. *Nature, Lond.,* **232**, 221.

BEECH, F. W. (1957). *The incidence and classification of cider yeasts.* Doctorial thesis. Bristol: University of Bristol.

BEECH, F. W. & CARR, J. G. (1960). Selective media for yeasts and bacteria in apple juices and ciders. *J. Sci. Fd Agric.,* **11**, 35.

BEECH, F. W. & DAVENPORT, R. R. (1969). The isolation of non-pathogenic yeasts. *Isolation Methods for Microbiologists.* Society of Applied Bacteriology, Technical series 3. (D. A. Simpson and G. W. Gould, eds). London and New York: Academic Press.

BEECH, F. W. & DAVENPORT, R. R. (1971). Isolation, purification and maintenance of yeasts. *Methods in Microbiology,* Vol. 4 (J. R. Norris, D. W. Ribbons and C. Booth, eds). London and New York: Academic Press.

BEECH, F. W., DAVENPORT, R. R., GOSWELL, R. W. & BURNETT, J. K. (1968). Two simplified schemes for identifying yeast cultures *In Identification Methods for Microbiologists.* Part B , No. 2, pp. 150–75. London and New York: Academic Press.

BORRADAILE, L. A., POTTS, F. A., EASTHAM, L. E. S. & SAUNDERS, J. T. (1959). Revised by G. A. Kerkut. *The Invertebrata.* Cambridge: Cambridge University Press.

BOWEN, J. F. (1962). *The ecology of cider yeasts.* Doctorial thesis. Bristol: University of Bristol.

CAMPBELL, I. (1971). Comparison of serological and physiological classification of the genus *Saccharomyces*. *J. gen. Microbiol.*, **63**, 189.

COPE, D. W. (1969). Soil survey of Long Ashton Research Station. *Rep. Long Ashton Res. Stn for 1968*, 170.

CUSTERS, T. J. (1940). *Onderzoekingen over het gistgeslacht* Brettanomyces. Doctorial thesis. Holland: Delft.

DAVENPORT, R. R. (1968). *The origin of cider yeasts*. Thesis. London: Institute of Biology.

DAVENPORT, R. R. (1970*a*). *Epiphytic yeasts associated with the developing grape vine*. M.Sc. thesis. Bristol: University of Bristol.

DAVENPORT, R. R. (1970*b*) *Rep. Long Ashton Res. Stn for 1969*, 136.

DAVENPORT, R. R. (1972). *Rep. Long Ashton Res. Stn for 1971*, 186.

FOWELL, R. R. (1952). Sodium acetate agar as a sporulation medium for yeasts. *Nature, Lond.*, **170**, 578.

KLEYN, J. G. (1954). A study of some environmental factors controlling sporulation using a new sporulation medium. *Wallerstein Labs Commun.*, **17**, 91.

KREGER VAN RIJ, N. J. W. (1964). *A taxonomic study of yeast genera* Endomycopsis, Pichia *and* Debaryomyces. Doctorial thesis. Holland: Leiden.

LODDER, J. (1970). *The Yeasts*. Amsterdam: North Holland Publishing Company.

LODDER, J. & KREGER VAN RIJ, N. J. W. (1952). *The Yeasts, a Taxonomic Study*. Amsterdam: North Holland Publishing Company.

LUND, A. (1956). Yeasts in nature. *Wallerstein Labs Commun.*, **19**, 221.

OLDROYD, H. (1958). *Collecting, Preserving and Studying Insects*. London: Hutchinson.

OXOID (1965). *Oxoid Manual* (3rd Ed.), London.

OXOID (1969). *Oxoid Manual* (3rd Ed.), (reprint), London.

PHAFF, H. J., MILLER, M. W. & MRAK, E. M. (1966). *The Life of Yeasts*. Cambridge, Massachusetts: Harvard University Press.

RACZ, J. & SLABA, K. (1971). Der allelopathische einflub der unkräuter im weinburg. *Mitt. Rebe u. Wein, Obst. Früchteverwert.*, **21**, 264.

RAW, F. (1959). Estimating earthworm populations by using formalin. *Nature, Lond.*, **184**, 1661.

SCAMONI, A. (1966). Biogenozönose-phytozönose Biosoziologie. *Bericht über das, internationale Symposium in Stolzenau/Weser 1960*. (R. Tuxen, ed.). The Hague Holland: Junk Publ.

TANSLEY, A. G. (1935). The use and abuse of vegetational concepts and terms. *Ecology*, **16**, 284.

VAN DYNE, G. M. (1969). Implementing the ecosystem concept in training in natural resource sciences. *The Ecosystem Concept in Natural Resource Management*. New York and London: Academic Press.

VAN DER WALT, J. P. (1952). *On the yeast* Candida pulcherrima *and its pigment*. Doctorial thesis. Holland: Delft.

WALTERS, L. S. (1943). Studies of yeasts causing defects in beers. *J. Inst. Brew.*, **49**, 245.

WICKERHAM, L. J. (1951). Taxonomy of yeasts. *Washington, DC US Dept Agric. Tech. Bull. No. 1029.*

WILES, A. E. (1953). Identification and significance of yeasts encountered in the brewery. *J. Inst. Brew.*, **59**, 265.

WINKLER, A. J. (1965). *General Viticulture*. California, USA: University of California Press.

The Estimation of Soil Protozoan Populations*

J. F. DARBYSHIRE

*Department of Microbiology, The Macaulay Institute for Soil Research,
Craigiebuckler, Aberdeen AB9 2QJ, Scotland*

The common methods of estimating soil protozoan populations are modifications of the dilution method used in soil bacteriology. Direct enumeration of soil protozoa amongst soil particles or in soil suspensions has usually
yielded much lower estimates than those obtained from the prolonged incubation of suitable dilutions of soil suspensions. Although Rahn (1913) and
Killer (1913) were the first investigators to count soil protozoa by a dilution
method, it was Cutler (1919, 1920), who first rigorously tested the dilution
method and developed the hydrochloric acid modification for estimating the
number of active protozoa in soil. Subsequently Cutler, Crump and Sandon (1922) used this dilution method to estimate the numbers of the
commonest soil protozoa for 365 consecutive days in the upper 9 in. of soil
from an experimental plot on Barnfield at Rothamsted Experimental
Station, Harpenden, England. This investigation was notable, not only for
its magnitude but because it contained the first estimates of soil protozoan
populations that had been subjected to statistical analysis. Singh (1946,
1955) increased the replication that was practically possible with Cutler's
dilution method by suggesting that all the replicate aliquots from one dilution should be put together into one Petri dish and separated from one
another by small glass rings set into the agar. Previously, the individual
aliquots of the soil dilutions were incubated in separate Petri dishes. As a
result of his studies on the feeding habits of soil protozoa (Singh, 1941a, b,
1942, 1945, 1946, 1947a, b, 1948; Anscombe and Singh, 1948), Singh concluded that the peptone nutrient agar used in Cutler's dilution method may
encourage the growth of inedible or toxic soil microflora, which would
hinder the full development of the soil protozoan population. Singh (1946,
1955) recommended the use of 0·5% (w/v) NaCl agar supplied with a
thick suspension of a pure culture of a suitable edible bacterium (e.g.
Aerobacter aerogenes) instead of the peptone nutrient agar. Since most
recent estimates of soil protozoan populations have been made with some

---

slight modification of Singh's so-called "ring method", an abbreviated version of his description (Singh, 1955) is quoted below.

Singh's Ring Method

" . . . About 15 ml of non-nutrient agar . . ." (i.e. 0·5% NaCl agar) " . . . is poured into each Petri dish and eight sterile glass rings (each 2 cm internal diameter, 1 cm depth and 1–2 mm thickness) are placed in it before the agar solidifies. A thick suspension of *Aerobacter* sp., strain 1912 (Singh 1941) or a mixture of edible species of bacteria (see Singh 1946) is spread over the agar surface in each ring as a circular patch or 'bacterial circle'. The plates are then ready for inoculation . . ." (Fig. 1). " . . . Twelve 6-inch borings of a field soil are taken and thoroughly mixed to form a composite sample which is then passed through a 3 mm sieve. Ten grammes of this soil are shaken for 5 minutes with a 0·5 per cent NaCl solution, giving a dilution of $\frac{1}{5}$. A series of twofold dilutions are then made, ranging from $\frac{1}{10} - \frac{1}{81,920}$. In order to make each dilution 5 ml of the next higher dilution is added to 5 ml of the 0·5 per cent NaCl solution contained in a test tube. Eight glass rings are then inoculated in the centres of the 'bacterial circles' with 0·05 ml of each of these dilutions.

After inoculating the 'bacterial circles' with different dilutions of soil, the plates are incubated for two weeks at 21–22°C without turning them over. In order to help in the development of flagellates and ciliates a drop of sterile

FIG. 1. Petri dish with 8 glass rings set in 0·5% NaCl agar. The inside of each ring has been inoculated with a thick suspension of bacteria as recommended by Singh (1955).

tapwater or 0·5 per cent NaCl solution is added to the cultures when the agar is dry. The cultures are examined after one and two weeks under the low power of a microscope for the presence or absence of Protozoa in each glass ring. The entire surface of each 'bacterial circle' can readily and rapidly be explored. Thus the error involved in sampling an entire plate surface is obviated and the examination is sufficient for detecting the presence or absence of amoebae, flagellates and ciliates.

In order to make quantitative studies of individual species, a sample from each of the cultures showing protozoal growth is examined under the high power of a microscope. The number of Protozoa per gramme of soil is calculated from the count of negative cultures (that is, showing no Protozoa) by applying Fisher's (1922) method for negative plates (see Fisher and Yates, 1943, Table VIII_2 and Appendix to Singh 1946) . . . Workers using this method are referred to a table given in the appendix to Singh's 1946 paper, which shows estimated numbers of Protozoa per gramme of soil. The 5 per cent significance level difference between two individual counts is eight cultures, corresponding to approximately 100 per cent difference in population estimates. The increase in replication greatly reduces the limits of uncertainty that attend the estimate. As higher replication has been shown to be practicable, eight replicates should be considered as minimum. . . .

. . . First a count of the total Protozoa (active + cystic) is made and then of the cysts by killing the active forms. The number of cysts subtracted from the total organisms give the active number of Protozoa per gramme of soil. . . . One sample of 10 grammes of soil is used to count total Protozoa by the ring method. Another sample of 10 grammes is then treated with 40–50 ml of 2 per cent HCl (the carbonate content of the soil is estimated and sufficient HCl is added to leave an excess of acid) over night. After removing most of the acid, 0·5 per cent NaCl solution is added to make a dilution of $\frac{1}{5}$. From this further dilutions are made and the numbers of cystic Protozoa determined. . . .''

The Modification of Singh's Ring Method in use at the Macaulay Institute

The modification in use at the Macaulay Institute requires fewer preparatory manipulations than the original method (Singh, 1955) and the investigator can concentrate more on microscopical observation.

Soils are sampled by one of two methods. In both methods the sampling instruments are sterilized before use.

(a) Borings are made at 2 m intervals with a soil auger on a line across the experimental area at the required depth to provide sufficient soil to fill a 1 lb Kilner jar. The next series of borings are made in the same way except that all the sites are situated 60 cm to the right of the corresponding borings in the previous series of samples. This is essentially the technique used by Cutler et al. (1922) and is the convenient method to use when the experimental area is small and a series of samples are required.

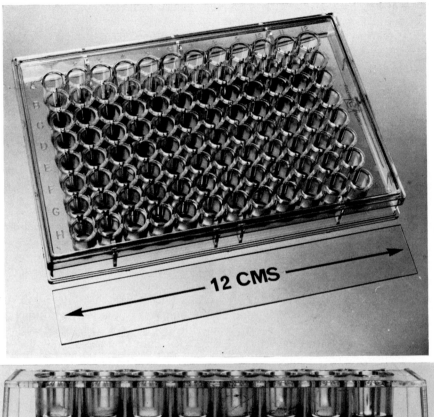

FIG. 2. Rigid styrene plates (top and side views) with 96 wells, arranged in 8 rows of 12. Each well has a working capacity of 0·125 ml.

(b) Up to 6 soil pits are dug at selected sites within the experimental area to the depth required to expose the soil horizons under investigation. The outer 5–6 cm of the relevant soil horizon is removed with a scalpel and then sufficient soil is collected with a cork borer (25 mm diam) from all the soil pits to fill a 1 lb Kilner jar. The second method of sampling is used when the experimental area is extensive and when particular soil horizons are

	Dilution A													Dilution B												
	1	2	3	4	5	6	7	8	9	10	11	12	Total	1	2	3	4	5	6	7	8	9	10	11	12	Total
COLPODA sp. (30μ long)	X	X			X			X		X	X	X	7	X	X				X			X				4
BALANTIOPHORUS sp.								X					1													
LIMAX AMOEBA (Eruptive Pseudopodia)	X		X	X	X		X	X	X	X		X	9	X	X	X	X	X	X	X	X	X	X	X		11
HARTMANNELID AMOEBAL CYST	X				X		X	X	X	X		X	7	X			X						X	X	X	5
TESTATE AMOEBAE		X		X	X	X	X					X	6										X			1
NUCLEARIA sp.	X			X				X		X			4										X			1
ACTINOPHRYS sp.			X			X		X		X			4			X		X					X	X		4
OIKOMONAS (2 spp.)	X			X	X	X		X	X	X	X	X	9	X	X	X	X				X	X		X	X	8
HETEROMITA sp.	X	X		X	X	X	X	X	X			X	10		X	X	X	X	X		X	X	X	X	X	10
SCYTOMONAS sp.		X			X					X			3		X				X		X		X			4
CERCOMONAS sp.			X		X			X	X	X	X		6		X	X	X			X		X	X			6
CERCOBODO sp.				X	X		X				X	X	5	X			X	X		X	X	X				6
MASTIGAMOEBA sp.																				X						1
ALLAS diplophysa								X					1													
UNKNOWN FLAGELLATE 1	X	X		X	X	X	X					X	8		X											1
UNKNOWN FLAGELLATE 2	X												1													
UNKNOWN FLAGELLATE 3							X	X					2						X							1
UNKNOWN CYST														X	X		X	X			X	X	X	X	X	9

FIG. 3. An example of the data sheet used for recording the presence or absence of protozoa in each aliquot of a soil dilution series.

under investigation. In the laboratory, the soil in each Kilner jar is thoroughly mixed before the larger particles are removed with a 150 mm diam sieve (3 mm mesh). The sieved soil samples are stored in closed Kilner jars until they are used for fresh weight determinations and the dilutions.

Soil-extract fluid is used as the diluent instead of saline as recommended by Singh. This extract is prepared from soil obtained from the Institute's garden. The soil is first air dried and then passed through a 200 mm diam sieve (4·76 mm mesh, Endecott Filters Ltd., London, England). One kg of sieved soil is mixed with 1 litre of tap water and autoclaved (121°/30 min). The supernatant soil suspension is decanted, filtered through filter candles (Size No. 2, British Berkefeld Ltd., Tonbridge, England) and finally autoclaved (121°/20 min).

The original dilution of 10 g (fresh weight) of soil in 50 ml sterile soil extract is shaken for 5 min on an orbital shaker (Model G.25, New Brunswick

Scientific Co., New Brunswick, USA) at 60 rev/min. Wide mouthed pipettes and 100 ml conical flasks are used to prepare the series of twofold dilutions. All this glassware is coated with a layer of silicone (Repelcote, Hopkins and Williams Ltd., Chadwell Heath, England). The pipettes are recalibrated before they are used. The 12 replicate aliquots (0·05 ml) of each dilution are pipetted into separate wells of sterile rigid styrene plates (Falcon Plastics Ltd., supplied by Becton Dickinson Ltd., Wembley, England) shown in Fig. 2. These plates are incubated at 21° for 4 weeks inside confectionary display trays with small beakers of water to reduce evaporation from the soil dilutions. At 4, 7, 10, 14, 21 and 28 days after inoculation each soil dilution is examined with an inverted microscope using long focus objectives with total magnifications up to ×800 and the presence or absence of protozoa noted. The microscopical observations are recorded on data sheets in the manner shown in Fig. 3.

TABLE 1. A comparison between the Singh (1955) method and the Macaulay modification for estimating the number of soil protozoa/g oven-dry soil

Organisms	Method	
	Singh	Macaulay
Ciliophora		
Colpoda sp	118	274*
Balantiophorus sp	0	108
Vorticella sp	0	115
Sarcodina		
Hartmanella rhysodes ⎫	3770	2603
Unidentified limax amoeba ⎭		
Nuclearia sp	0	136
Actinophrys sp	0	194*
Testate amoebae	0	154*
Mastigophora		
Oikomonas 2 spp	205	690*
Heteromita globosa	333	2761*
Scytomonas pusilla	0	183*
Cercomonas 2 spp	439	387
Phalansterium solitarum	0	145
Cercobodo sp	0	258*
Allas diplophysa	0	108
Unidentified flagellate sp 1	0	205*
Unidentified flagellate sp 2	0	108
Unidentified flagellate sp 3	0	122

The soil was oven dried at 105° for 24 h. The figures marked with an asterisk* are significantly larger than the equivalent numbers obtained by the Singh method at the 0·05 level of probability.

A comparison between the Singh (1955) method and the Macaulay modification of his method has shown that significantly larger populations of flagellates and ciliates can be obtained with the Macaulay modification (Table 1). One disadvantage of this modification is that some of the most concentrated soil dilutions may be opaque in the inverted microscope. This problem can be reduced by lightly pushing the soil sediment to one side of the well with a small sterile glass rod.

The correct identification of protozoa in the soil dilutions usually requires some reference to studies of protozoan systematics. A short list of relevant books and scientific papers is given in the Appendix. Some reagents and stains which may aid identification are also listed in the Appendix.

Future Possible Improvements to Singh's Ring Method

At present, 2 instruments (Cooke Engineering Co., Alexandria, Virginia, USA supplied by Flow Laboratories Ltd., Irvine, Scotland), which may further reduce the manual operations involved in the series of twofold dilutions (Figs 4–7), are being tested. Originally, these instruments were developed for viral serological investigations but preliminary work suggests that they can also be used to produce satisfactory twofold dilutions for the estimation of soil protozoan populations. The multi-channel automatic

FIG. 4. Multi-channel automatic pipetter.

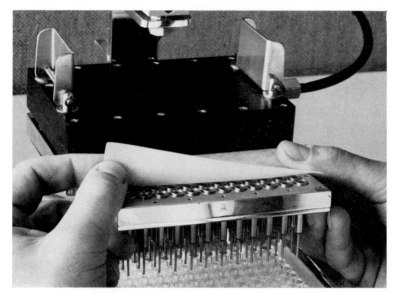

FIG. 5. Detail of multi-channel automatic pipetter.

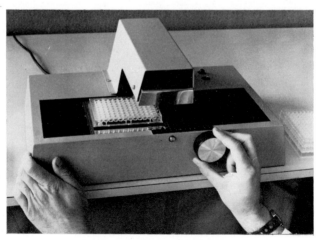

FIG. 6. Automatic diluter.

pipetter (Fig. 4) can be used to deliver 25 μl of soil extract fluid into all the wells of the rigid styrene plates (Fig. 2), except the wells in the first row, which are filled manually with 50 μl aliquots of the 1/320 dilution of the original soil suspension. The automatic diluter (Fig. 6) fitted with 25 μl microdiluters can then be used to prepare a twofold dilution series by transferring 25 μl aliquots from the wells in one row to corresponding wells in the next row.

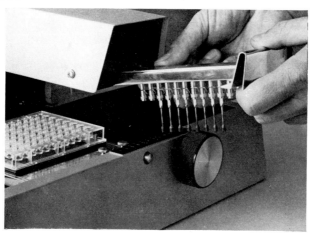

FIG. 7. Detail of automatic diluter.

Dixon (1937) commented that if soil extract is heated for a long period, it has a depressant effect on the subsequent cultivation of protozoa. Echlin (1968) similarly concluded that the duration of autoclaving, conventionally used by bacteriologists (121°/15 min), induced degradative changes in the soil extract fluid. Echlin recommended that the soil extract fluid should be autoclaved at 121°/3 min when used in culture media for blue-green algae. It would seem probable, therefore, that larger estimates of soil protozoan populations would be obtained if the soil extract was sterilized by filtration. A cool soil extraction would possibly be a further improvement. Grunda (1970) found that humic and fulvic acids, which were extracted from a chernozem of a humus podzol and added to nutrient agar, increased microfloral counts. The maximum increase occurred at lower concentrations of fulvic acids (20–100 mg C/100 ml agar) than with humic acids (100–400 mg C/100 ml agar) from the humus podzol. He also found that humic acids extracted from the podzol and the chernozem had different effects on microfloral growth. Protozoa may also react differently to soil extracts from different soils and to different fractions from the same soil extract. The results of experiments designed to test these hypotheses and possible improvements to Singh's Ring method will be published elsewhere.

184 J. F. DARBYSHIRE

Acknowledgement

Flow Laboratories Ltd., Irvine, Scotland are thanked for providing the photographs of the automatic diluter and automatic pipetter.

References

ANSCOMBE, F. J. & SINGH, B. N. (1948). Limitation of bacteria by micro-predators in soil. *Nature, Lond.*, **161**, 140.

CUTLER, D. W. (1919). Observations on soil protozoa. *J. agric. Sci. Camb.*, **9**, 430.

CUTLER, D. W. (1920). A method for estimating the number of active protozoa in soil. *J. agric. Sci. Camb.*, **10**, 135.

CUTLER, D. W., CRUMP, L. M. & SANDON, H. (1922). A quantitative investigation of the bacterial and protozoan population of the soil, with an account of the protozoan fauna. *Phil. Trans. R. Soc., B.* **211**, 317.

DIXON, A. (1937). Soil protozoa; their growth on various media. *Ann. appl. Biol.*, **24**, 442.

ECHLIN, P. (1968). The culture of the blue-green algae and the use of the electron microscope in identification and classification. In *Identification Methods for Microbiologists Part B* (B. M. Gibbs and D. A. Shapton, eds). London and New York: Academic Press.

FISHER, R. A. (1922). On the mathematical foundations of statistics. *Phil. Trans. R. Soc., A.* **222**, 309.

FISHER, R. A. & YATES, F. (1943). *Statistical tables for biological, agricultural and medical research* (2nd Ed.) London and Edinburgh: Oliver & Boyd.

GRUNDA, H. (1970). Der Einfluss der Humusstoffe auf die Anzahl der Boden-kleinlebewesen in Kulturen. *Zentbl. Bakt. ParasitKde. Abt. II*, **125**, 584.

KILLER, J. (1913). Die Zählung der Protozoën im Boden. *Zentbl. Bakt. ParasitKde, Abt, II*, **37**, 521.

RAHN, O. (1913). Methode zur Schätzung der Anzahl von Protozoën im Boden. *Zentbl. Bakt. ParasitKde, Abt. II*, **36**, 419.

SINGH, B. N. (1941a). Selectivity in bacterial food by soil amoebae in pure mixed cultures and in sterilized soil. *Ann. appl. Biol.*, **28**, 52.

SINGH, B. N. (1941b). The influence of different bacterial food supplies on the rate of reproduction in *Colpoda steinii*, and the factors influencing encystation. *Ann. appl. Biol.*, **28**, 65.

SINGH, B. N. (1942). Selection of bacterial food by soil flagellates and amoebae. *Ann. appl. Biol.*, **29**, 18.

SINGH, B. N. (1945). The selection of bacterial food by soil amoebae, and the toxic effects of bacterial pigments and other products on soil Protozoa. *Br. J. exp. Path.*, **26**, 316.

SINGH, B. N. (1946). A method of estimating the numbers of soil Protozoa, especially amoebae, based on their differential feeding on bacteria. *Ann. appl. Biol.*, **33**, 112.

SINGH, B. N. (1947a). Studies on soil Acrasieae. 1. Distribution of species of *Dictyostelium* in soils of Great Britain and the effect of bacteria on their development. *J. gen. Microbiol.*, **1**, 11.

SINGH, B. N. (1947b). Studies on soil Acrasieae. 2. The active life of species of

Dictyostelium in soil and the influence thereon of soil moisture and bacterial food. *J. gen. Microbiol.*, **1**, 361.

SINGH, B. N. (1948). Studies on giant amoeboid organisms. 1. The distribution of *Leptomyxa reticulata* Goodey in soils of Great Britain and the effect of bacterial food on growth and cyst formation. *J. gen. Microbiol.*, **2**, 8.

SINGH, B. N. (1955). Culturing soil protozoa and estimating their numbers in soil. In *Soil Zoology* (D. K. McE. Kevan, ed.). London: Butterworths Scientific Publications.

Appendix

Useful reagents and stains

Schaudinn's Fluid

This is a good fixative for many protozoa.
Saturated aqueous mercuric chloride, 2 parts
Absolute ethanol, 1 part
Some glacial acetic acid (up to 5 % v/v) is added to the mixture just before use.

Osmium Tetroxide

Many delicate protozoa can be fixed after a few minutes exposure to the vapour from aqueous solution of OsO_4 (w/v) 1 or 2 %. As this is a dangerous vapour which can cause conjunctivitis, all the operations should be performed in a fume cupboard and eye goggles should be worn.

Nickel sulphate

This is a good anaesthetic for motile protozoa. A small drop of $NiSO_4$ (1 % w/v) solution can be mixed with an equal volume of the protozoan culture fluid. The speed of movement of the protozoan will be gradually slowed down without any apparent body deformation.

Mayer's albumen

This is often used to stick protozoa to microscope slides. Beat the white of one egg and transfer it to a beaker to allow the suspended matter to rise to the surface. After this surface material has been skimmed off, the albumen is diluted with an equal volume of glycerine. Approximately 2 g of sodium salicylate or thymol is also added to prevent rapid microbial decomposition. The protozoa can be attached to very thin smears of egg albumen on microscope slides.

Lugol's iodine

This is a useful reagent for staining flagella or cilia. Potassium iodide (6 g) is first dissolved in *c.* 20 ml of glass distilled water. Iodine (4 g) is then added to the KI solution and volume is made up to 100 ml with glass distilled water.

Neutral red

This is a useful vital stain for demonstrating vacuoles when aqueous solutions at concentrations between 1:10,000 and 20,000 (w/v) are used.

Iron haematoxylin

This is a good nuclear stain for many protozoa. The slides with the attached protozoa are washed first in ethanol (70 % v/v), then immersed in ferric alum mordant solution (1 % (w/v) in ethanol 70 % (v/v) for 10 min and washed in ethanol (70 % v/v). Subsequently, the slides are stained in haematoxylin (1 % w/v) for 10 min and washed again in ethanol (70 % v/v), before they are destained under a microscope with the mordant. The slides can then be dehydrated in ethanol (90 % v/v and absolute), cleared in xylene or cedar wood oil and finally mounted in Canada balsam or Euparol. The cytoplasm can be stained with Orange G (saturated solution in ethanol 90 % (v/v)).

Feulgen method of staining protozoan nuclei

The Feulgen reaction depends on the hydrolysis of DNA with the release of aldehydes, which are stained pink by leuco-basic fuchsin (Schiff's Reagent).

Staining sequence

(1) Fix the slides and attached protozoa with corrosive sublimate (2–20 min). Corrosive sublimate contains 98 parts $HgCl_2$ 6 % (w/v) and 2 parts glacial acetic acid.

(2) Wash the slides for a short time (actual duration should be the same as the period of fixation and not over 20 min).

(3) Wash the slides in distilled water for 2 min.

(4) Wash the slides in 1 N-HCl at room temperature for 2 min.

(5) Wash the slides in 1 N-HCl at 60° for 2 min.

(6) Rinse the slides in 1 N-HCl at room temperature.

(7) Rinse the slides in distilled H_2O.

(8) Wash the slides in sulphurous acid for 2 min. Sulphurous acid rinses: add 15 ml of potassium metabisulphite solution 10 % (w/v) and 15 ml of 1 N-HCl to 270 ml of distilled H_2O. This sulphurous acid solution should be prepared on the day it is required.

(9) Stain the slides in leuco-basic fuchsin for $1\frac{1}{2}$–2 h.

(10) Rinse the slides in sulphurous acid baths for sufficient time to remove all the free unreacted leuco-basic fuchsin. Usually 2 or 3 quick changes are used.

(11) Wash in tap water for 10–15 min.

Light green/picric acid (light green, 1 g, and picric acid, 0·5 g, in 100 ml of ethanol 90 % (v/v)) may be used as a counterstain for 10 sec.

(12) Rinse the slides in tap water.

(13) Dehydrate the slides with ethanol (50 %, 70 %, 90 %, 95 % v/v, absolute), clear in xylene and mount in Canada balsam.

Schiff's reagent

Dissolve basic fuchsin (1 g) in 200 ml of boiling distilled water in a stoppered 1 litre flask and shake for 5 min. Cool the solution to exactly 50°, filter and add 20 ml of 1 N-HCl to the filtrate. Cool the filtrate further to 25° and add 1 g of potassium metabisulphite. Store the filtrate in the dark for 18–24 h, add 2 g activated charcoal and shake the mixture for 1 min. Filter the mixture to remove the charcoal and store the filtrate in the dark at 0–4°.

Chatton-Lwoff silver line stain for ciliates

Staining sequence

(1) Fix the ciliates in Champy's fluid for 1–3 min. Champy's fluid contains: 20 ml osmium tetroxide (2% w/v), 35 ml chromic acid (1% w/v) and 35 ml potassium dichromate (3% w/v). The chromic acid solution can be prepared by adding 1 g of chromium trioxide crystals to 100 ml distilled water.

(2) Wash the ciliates twice in Da Fano's fixative and then leave in a third change for 2 to 3 h. The ciliates can be stored in this fixative for several weeks. Da Fano's fixative contains: cobalt nitrate, 1 g; sodium chloride, 1 g; formalin, 10 ml, and distilled water, 90 ml.

(3) Concentrate the ciliates by centrifugation (5 min at 400 g) after fixation in Champy's and Da Fano's fluids. Then place one drop of concentrated ciliates on a warm slide. Add 1 drop of warm (35–40°) saline gelatine and mix with tip of a warm needle. Draw off the excess fluid so as to leave the specimens grouped in a thin layer of gelatine. (Saline gelatine: powdered gelatine, 10 g; sodium chloride, 0·05 g, and distilled water, 100 ml; store in a refrigerator until required.)

(4) Place the slide with the attached ciliates in a Petri dish with wet filter paper. Cool the Petri dish either in a refrigerator or on ice for *c.* 2 min until the gelatine has set. Add cold (5–10°) 3% (w/v) silver nitrate to the slide and keep it in the dark for 10–20 min.

(5) Flush the slide with cold distilled water and place it in a white dish containing cold distilled water to a depth of 3–4 cm. Expose the slide to light for 20–30 min. The preparation must be kept cold.

(6) Wash the slide in ethanol 70% (v/v), dehydrate as in Feulgen stain method in ethanol (90, 95% v/v) and absolute ethanol, clear in xylene, and mount in Canada balsam. Any preparation, which is too black at the end of stage (4) may be toned down under a microscope with a drop of gold chloride 1% (w/v). The toning can be stopped instantly by plunging the slide into a large beaker of cold distilled water at the desired time.

References to protozoan systematics

CAULLERY, M., CHATTON, E., CUÉNOT, L., DEFLANDRE, G., GRASSÉ, P-P., HOLLANDE, A., LE CALVEZ, J., PAVILLARD, J., POISSON, R. & TRÉGOUBOFF, G. (1952, 1953). *Traité de Zoologie.* Anatomie, Systématique, Biologie (P-P. Grassé, ed.). 1, fasc. 1 and 2. Paris: Masson and Cie.

CORLISS, J. O. (1961). *The Ciliated Protozoa: Characterization, Classification, and Guide to the Literature.* London and New York: Pergamon Press.

GRANDORI, R. & GRANDORI, L. (1934). Studî sur Protozoi del terreno. *Annali R. Ist. sup. agrar. Milano,* **1,** 1.

HONIGBERG, B. M., BALAMUTH, W., BOVEE, E. C., CORLISS, J. O., GODJICS, M., HALL, R. P., KUDO, R. R., LEVINE, N. D., LOEBLICH, A. R., WEISER, J., WEINRICH, D. H. (1964). A Revised Classification of the Phylum Protozoa. *J. Protozool.,* **11,** 7.

KAHL, A. (1930–35). Urtiere oder Protozoa I: Wimpertier oder Ciliata (Infusoria) eine Bearbeitung der freilebenden und ectocommensalen Infusorien der Erde, unter Ausschlus der marinen Tintinnidae. In *Die Tierwelt Deutschlands* (F. Dahl, ed.). Vols 18, 21, 25, 30. Jena: G. Fisher.

PAGE, F. C. (1967a). Taxonomic Criteria for Limax Amoebae, with Descriptions of 3 New Species of Hartmanella and 3 of Vahlkampfia. *J. Protozool.*, **14,** 499.

PAGE, F. C. (1967b). Re-Definition of the Genus *Acanthamoeba* with Descriptions of Three Species. *J. Protozool.* **14,** 709.

SANDON, H. (1927). *The Composition and Distribution of the Protozoan Fauna of the Soil.* Edinburgh and London: Oliver and Boyd.

SINGH, B. N. (1952). Nuclear division in nine species of small free-living amoebae and its bearing on the classification of order Amoebida. *Phil. Trans. R. Soc., B.,* **236,** 405.

SINGH, B. N. & DAS, S. R. (1970). Studies on pathogenic and non-pathogenic small free-living amoebae and the bearing of nuclear division on the classification of the order Amoebida. *Phil. Trans. R. Soc., B.* **259,** 435.

Sampling of Pharmaceuticals for Microbial Contamination

M. C. ALLWOOD

*Department of Pharmacy, University of Manchester,
Manchester, England*

The microbial contamination of non-sterile pharmaceutical products was highlighted in 1966 (Kallings, Ringertz and Silverstolpe, 1966). A number of studies have since been reported illustrating the wide range of products which, from time to time, contain viable microorganisms. The total number of detectable living cells may vary from less than 100 to an excess of 10^6 colony forming units/g. The type of product appears to be little respected by microorganisms and high counts have been observed in such widely different medications as aqueous mixtures, creams and tablets. In consequence, in describing methods for sampling pharmaceuticals for microbial contamination, each type of product must be considered separately.

Although various methods have been described for the isolation of microorganisms from different product forms, there is a paucity of critical comparative data to suggest the relative merits of each procedure. Consequently, those described in this paper arise from methods employed by various authors supplemented with an appraisal of the merits of the technique where data are available. It should also be stressed that, since most reports are concerned with the "natural" contamination of products obtained from in-use sources, the proportion of contaminants recovered is not measurable and remains a matter of conjecture.

Sampling Techniques

Before describing various methods for sampling different drug forms, some general points should be stressed. Firstly, care is required when taking the sample to ensure that it is representative of the product under examination. Adequate stirring is necessary and account should be made of the effects of the cap liner in supporting microbial growth. The method of assessing numbers of viable microorganisms must be related to various factors in the sampling procedure. Methods requiring the sample to be in an aqueous non-toxic liquid, such as tube dilution and solid media counts, should be distinguished from a filtration method for assessing viability in which the sample must be dissolved in a suitable solvent. Adequate recovery of viable cells from the pharmaceutical preparation may be encouraged by the use of a complex organic menstruum (such as nutrient broth or peptone water— Oxoid—1% w/v), glass beads, a small concentration of a surface-active agent and sufficient agitation by shaking or homogenization.

TABLE 1. Minimum detectable viable counts of different products

Prepara-tion	Method	Sample size	Diluent or solvent	Dispersion or solu-tion time	Minimum dilution*	Minimum detectable viable count†
Solutions	Aqueous dilution	1 ml	Peptone water	1 min	10*	100/ml
Syrups, linctuses	Filtration	1 ml	Peptone water	1 min		10/ml
Mixtures	Aqueous dilution	1 ml	Peptone water + 0·1% v/v Tween 80	4–5 min	10*	100/ml
Tablets, capsules	Aqueous dilution	2 units	Peptone water + 0·1% v/v Tween 80	up to 60 min	10*	50/unit
Supposi-tories	Aqueous dilution	1–2	Peptone water + 0·1 v/v Tween 80	5 min	10*	50/unit
Creams, ointments	Aqueous dilution	1 g	Peptone water + 1% v/v Tween 80	5 min	10†	1000/g
	Filtration	0·5 g	20 ml IPM	1 min		100/g
		0·5 g	20 ml 1% v/v Tween 80	1 min (homo-genized)		100/g

* See text for details. †Where appropriate 1 ml samples added to recovery media and assuming a minimum detectable viable count of 10 colonies/agar plate. IPM, isopropyl myristate.

Aqueous solutions

Such preparations are readily sampled by pipetting a known volume. Viscous preparations, such as syrups and linctures, require dilution in a suitable diluent if plate counts are to be made. Viscous liquids are usually filterable without dilution (Hirsch, Canada and Randall, 1969). The filter pad should be washed after filtration with peptone water or a nutrient broth to remove residual material.

Suspensions

Suspensions are aqueous preparations containing suspended solids, usually in a high proportion (10–45%, w/v). The solids often sediment rapidly. Thorough shaking is essential immediately before removing a known volume in a pipette. Most suspensions readily pass through the tip of the pipette although retention of large particles may occur at the external orifice. If very large particles are present in a preparation, the material should be shaken with 10% (w/v) glass beads (diam, 3 mm) before sampling. Dilution is essential before counting because of the large amount of solid matter present, which may encumber growth and make the counting of colonies on solid media difficult. One ml should be added to 9 ml peptone water containing 0·5 g glass beads (diam,3 mm). Dispersion may be improved by the addition of 0·1% (v/v) Tween 80 (Robinson, 1971). Suspensions cannot usually be examined by a filtration method.

Tablets and Capsules

Pulverization of tablets has been recommended before sampling (Pederson and Szabo, 1968). However, a limited survey of naturally contaminated tablets suggests that such a procedure is unnecessary (Allwood, 1971) and the process of tablet disintegration in aqueous diluent is adequate. The tablets should be added to 10 ml peptone water containing 0·5 g glass beads and shaken vigorously until disintegrated. To allow for ease of counting in the recovery medium, the maximum amount of insoluble material should not exceed c. 5% (w/v). Capsules readily break apart to release the contents after shaking in peptone water, although the capsule itself will not disintegrate.

The method described will require adaptation to sample special forms of tablets such as enteric coated or sustained-release preparations. The ease of disintegration in diluent will depend on the coating, the presence of lipid material and the hardness of the tablet. Pulverization is often required and a concentration of surfactant higher than 0·1% (v/v) Tween 80 may be necessary if lipid material is present.

Suppositories

Glycerine suppositories will dissolve in 10 ml peptone water at 37° and 1% (v/v) Tween 80 will aid dispersion. However, further dilution may be

lows: 0·5 g of material is added to 20 ml filter-sterilized IPM pre-warmed to 37°. After mixing by shaking or homogenization, to dissolve the product, up to 4 ml is added to 100 ml peptone water containing 0·1% (v/v) Tween 80 pre-warmed to 37°. The maximum quantity added to this rinse solution is limited by the amount of lipid material present, since large amounts of fatty material greatly reduce the filtration rate. It should be noted that water present in the product is immiscible with IPM and therefore is suspended in the solvent, thus the samples for filtration should be removed immediately on completion of the mixing stage. After shaking, the mixture is filtered through a 0·45 μm membrane filter, and the filter washed with 100 ml peptone water. Tsuji *et al.* (1971) suggest that the ointment be dissolved in 50 ml IPM at 42° and immediately filtered. This alternative procedure allows for a greater quantity of product to be filtered. However, since IPM is toxic to many vegetative microorganisms, a compromise is recommended in measuring contamination of non-sterile pharmaceuticals. The method described will allow for the detection of 100 viable cells/g, while reducing the toxic effects of IPM to a minimum. Deliberate contamination of white soft paraffin has shown this method to allow at least 90% recovery of *B. megaterium* spores. The filtration method could also be used with Tween 80 as dispersing agent, which produced 45% recovery of spores from the aqueous layer after homogenization.

Comparing methods 1 and 2, it would appear that the latter technique offers greater opportunity to detect low levels of contamination, provided the contaminants are not excessively sensitive to IPM. Aqueous dispersion followed by plating out into a solid recovery medium will only detect levels of contamination of not less than 1000 organisms/g since the Tween 80 must be diluted to allow growth of viable cells (Table 1). With the information available, the use of the IPM–filtration method is the more suitable technique for examining creams and ointments, particularly when the product contains large proportions of heavy fats. Alternatively, homogenization in Tween 80 followed by filtration of the aqueous layer should allow for reasonable degrees of recovery with reduced toxicity.

Table 1 summarizes the sampling of various drug forms and suggests minimum detectable numbers of viable microorganisms, assuming 100% recovery of the viable cells from the product, 1% (w/v) peptone water or diluent may be replaced by other complex organic solutions. Nutrient broth, Brain Heart infusion broth and others have been used. The size of the sample may be governed by some factors of the sampling procedure, such as the proportion of insoluble material or amount of oils and fats present, but it would appear to be significant to relate the sample amount to the dosage quantity.

Summary

Standards for the microbial cleanliness of non-sterile pharmaceuticals are being implemented in many countries. A commonly quoted limit of 100 viable microorganisms/ml or g is already officially accepted in Sweden. It would appear that a major difficulty in implementing these standards is the detection method employed to analyse products for the presence of viable cells. Detection methods must be improved in many instances to allow for the isolation of small numbers of contaminants. For routine laboratory purposes, these procedures will preferably be automated to the fullest extent since regular testing during production should become normal practice. Such requirements will necessitate considerable effort to improve sampling and isolation procedures before a more acceptable degree of reliability is attained.

Acknowledgements

I thank Professor A. M. Cook for his support and Dr R. Hambleton for his help in carrying out this work.

References

ALLWOOD, M. C. (1971). Microbials in non-sterile drugs. *Manuf. Chem. Aer. News*, **42**, 50.

HIRSCH, J. I., CANADA, A. T. & RANDALL, E. L. (1969). Microbial contamination of oral liquid preparations. *Amer. J. Hosp. Pharm.*, **26**, 624.

KALLINGS, L. O., RINGERTZ, O. & SILVERSTOLPE, L. (1966). Microbial contamination of medical preparations. *Acta. Pharm. Suecica*, **3**, 219.

PEDERSON, E. A. & SZABO, L. (1968). Microbial content of non-sterile pharmaceuticals. II. Methods. *Dansk. Tidss. Farm.*, **42**, 50.

ROBINSON, E. P. (1971). *Pseudomonas aeruginosa* contamination of liquid antacids: A survey. *J. Pharm. Sci.*, **60**, 604.

TSUJI, K., STAPERT, E. M., ROBERTSON, J. H. & WAIYAKI, P. M. (1971). Sterility test method for petrolatum-based ointments. *Appl. Microbiol.*, **20**, 798.

Automation and Instrumentation Developments for the Bacteriology Laboratory

A. N. SHARPE

*Unilever Research Laboratory Colworth/Welwyn,
Colworth House, Sharnbrook, Bedford, England*

The four exhibits shown at the Society's Autumn Meeting demonstrated the development of lines of attack on problems of quality control bacteriology in the food industry taken at Colworth. The major problems to be overcome in this type of bacteriology were seen to be the lengthy analytical times imposed by conventional incubation periods, high material costs, and severe technical and service labour requirements of conventional bacteriological methods limiting the numbers of analyses which could be made. At the outset, faced with an almost complete absence of automation or instrumentation for the bacteriology laboratory it was evident that we would have to evolve our own solutions to many of the problems. Three main approaches were open to us. These were the biochemical detection of bacteria or their products, methodological innovations, particularly the development of "Stomacher" homogenizers and aids to the rapid Droplet Technique of counting bacteria, and automation of large portions of conventional counting methods. The various developments in each of the approaches are described in this chapter.

The Biochemical Approach

Development of new or improved biochemical methods of detecting bacteria has the greatest ultimate potential, since it is here that really rapid analytical procedures are likely to be found, capable of raising bacteriology to the level of a chemical or biochemical auto-analytical procedure. During a survey of the existing literature many more or less relevant contemporary techniques, such as chromatography (particularly pyrolysis GLC), infra-red and mass spectrometry, electrophoresis, electronic particle counting, dye reduction, ^{32}P absorption, $^{14}CO_2$ release and diaminopimelic acid determination were examined closely for application to quality control bacteriology but were found to be wanting in sensitivity, selectivity, freedom

H

from interference by food debris, or general practicability. Some of our own experimental work led to radiochemical methods of detecting *Staphylococcus aureus* through deoxyribonuclease activity (Sharpe and Woodrow, 1971) and coliform organisms through formic dehydrogenase activity. Despite the fundamental sensitivity of radiochemical analyses, a practical lower limit of detectability of *c*. 10^4 organisms was found for both these tests. Our general feeling at the time was that none of the biochemical methods then available was likely to lead to significantly greater sensitivity than this, once it was brought down to the practicality of detecting bacteria in foods.

Detection of bacteria through adenosine triphosphate (ATP) determination had been of obvious interest from the start because of the low concentration at which this compound can be detected and its ubiquity in living cells. An instrument specifically designed for microbiological ATP analyses has recently been brought on to the market, and we were grateful to the du Pont Company (UK) Ltd. (Wilbury House, Wilbury Way, Hitchin, Herts) for loaning us the first Luminescence Biometer to reach the UK, on which an evaluation of its potential for food bacteriology was made. The instrument is shown in Fig. 1.

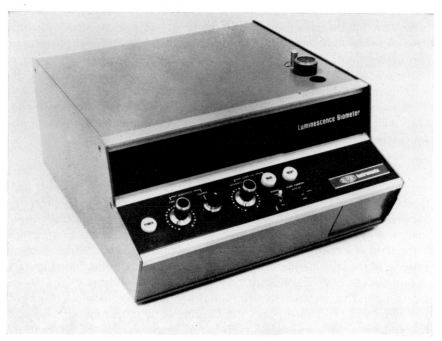

FIG. 1. The du Pont Luminescence Biometer.

Determination of bacterial ATP in foods

The Luminescence Biometer was first advertised in the USA in 1967 with claims that 1000 bacterial cells could be measured in 10 min. The instrument is basically a sensitive photometer with digital readout. The reaction used to determine biomass was well known before this date (McElroy, 1947; Addanki, Sotos and Rearick, 1966) and the du Pont achievement (in conjunction with Hazelton Laboratories Inc., Vancouver) was in making a suitable instrument and the rather delicate reagents available commercially. The biomass determination is based on the fact that all known terrestrial life is intimately associated with and dependent upon the nucleoside phosphate, ATP. Changes in the metabolic integrity of a living organism are accompanied by variations in the steady state ATP concentration of the organism. In a "dead" organism, i.e. one in which ATP production has ceased, the concentration of ATP decreases as it is consumed in phosphorylation reactions. The inference is that, for bacteriological purposes, the substrate for the bacteria, being dead, contains no ATP whilst the bacteria, being alive, contain ATP which can be measured and therefore used to indicate the level of contamination. In practice there are interpretative difficulties which are discussed later.

The low levels of ATP are determined by a bioluminescent reaction using an oxidizable substrate (luciferin) catalysed by an enzyme (luciferase), both components of which are obtained from firefly lanterns. Luciferin must first react with ATP before it can be oxidized, with light production, the two most important reaction steps being:

$$(1)\ E + ATP + LH_2 \underset{pH\ 7\cdot4}{\overset{Mg^{++}}{\rightleftharpoons}} E - LH - AMP + PP$$

$$(2)\ E - LH - AMP + O_2 \longrightarrow E - L - AMP + H_2O_2 + Light$$

E, enzyme (luciferase); LH, luciferin; PP, pyrophosphate, and AMP, adenylic acid. Sufficient O_2 is usually present in solution.

With a considerable excess of luciferin and luciferase, the maximum intensity of light is proportional to ATP concentration. The basic procedure is to inject cellular extracts containing ATP into cuvettes containing luciferin, luciferase and magnesium ion at pH 7·4. The light emitted is detected in the Luminescence Biometer by a photomultiplier tube and converted to an electric current which is measured and displayed. Figure 2 shows the form of the light emission curve. The reaction is specific for ATP.

As the concentration of ATP in living organisms is defined by fairly narrow limits the determination leads to a value of *biomass*, and this must be related to cell numbers through a factor which has to be measured for the particular organism concerned. Difficulties arise when mixed populations of cells occur (e.g. blood, pus, bacteria, yeasts, protozoa, etc.) because of the enormous range of relative cell masses. One can also see that the utility of the method for the determination of say, bacteria in foods, depends heavily on the assumption that dead cells contain no ATP. This is particularly evident when one considers that a fairly heavily contaminated food may contain no more than a millionth part, by weight, of bacteria. The natural level of ATP in the original tissue must fall by a very large factor before it becomes insignificant in terms of bacterial numbers.

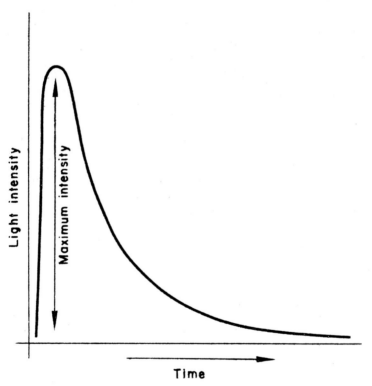

FIG. 2. A schematic representation of the bioluminescent reaction.

Using the Luminescence Biometer

Reagents are supplied by the manufacturers. Vials of the luciferin/luciferase mixture must be stored at −20° until required. Before commencing ATP determinations the following reagents must be made up.

Low response (LR) water. Distilled water is acidified with hydrochloric acid, boiled (5 min) and neutralized with sodium hydroxide. This can be autoclaved and stored.

Buffer solution. Morpholinopropane sulphonic acid (MOPS; 0·01 M, pH 7·4) is supplied with the instrument, although Tris can be used instead with excellent results. Freshly prepared LR water must be used, and 0·01 M–magnesium sulphate included.

Luciferin/luciferase (L/L). LR buffer (3 ml) are added to one vial of the enzyme/substrate mixture about 30 min before use. The solution is dispensed in 0·1 ml amounts in clean cuvettes. The time delay allows intrinsic luminescence to decay to a low level.

n-*Butanol.* Normal GPR reagent is satisfactory.

n-*Octanol.* GPR octanol should be washed with about 20 times its volume of distilled water and used without drying.

ATP solution. A standard solution of ATP at 10^{-7} g/ml should be prepared in LR buffer each day for calibration purposes.

ATP extraction. Two methods, with and without filtration, are recommended by the manufacturers. Whenever possible the filtration method should be used, but many food suspensions are quite unfilterable. Sample suspensions may be prepared in any way, but relatively non-destructive methods, which remove only surface contaminants (e.g. ultrasound) would be preferable.

Method 1—with filtration

A 25 mm × 0·45 μ membrane filter is washed with LR buffer and 1·0 ml of the cell suspension is filtered through and washed with 1·0 ml of buffer. The vacuum is broken and a clean centrifuge tube placed inside the vacuum flask.

Butanol (0·2 ml) is added to the filter, allowed to stand 20 sec to damage the cells and pulled through into the centrifuge tube. The process is repeated with a further 0·1 ml butanol. ATP is then washed through with 1·0 ml of LR buffer, the stand shaken to dislodge droplets and the vacuum broken again.

Octanol (8 ml) is added, the centrifuge tube stoppered and shaken vigorously. The two layers are separated by centrifugation, the top octanol/butanol layer being removed and 0·01 ml amounts of the aqueous layer used for ATP assay. The method given by the manufacturers includes a volumetric measurement at this stage as an aid to greater accuracy—it can often be omitted.

Method 2—without filtration

The sample suspension (1·0 ml) is placed directly in the centrifuge tube and 1·0 ml butanol added. After shaking 10 sec, 8 ml octanol are also added. The procedure then follows the method described above.

ATP assay

The instrument is first calibrated using the standard ATP solution, and its internal standard (light source) adjusted to equal this value. The sensitivity of the instrument is then adjusted so that the read-out equals the standard ATP concentration. The method of injecting the sample using an automatic syringe, and taking the reading are straightforward.

Before use with unknown ATP extracts the automatic syringe should be washed 3 times in 2 N–hydrochloric acid, and 3 times in LR buffer. The extract (0·01 ml) is then drawn into the syringe, and injected into a fresh cuvette of luciferin/luciferase as soon as this is in place in the light-tight reaction compartment. The machine's controls ensure adequate time elapsing for the maximum light intensity to be reached before the read-out is activated. Duplicate assays should be made on most solutions. Frequent checks on the instrument sensitivity, and on the background levels of the reagents should also be made.

Typical ATP values of microorganisms and foods

The ATP content of bacterial cells varies according to their size and metabolic state. Values between 0·2–1·0 fg (1 femtogram $= 10^{-15}$ g) ATP can be expected. Bacterial spores contain smaller but still detectable quantities of ATP. Yeast cells may contain 2–500 times this amount, and other cells (protozoa, leucocytes, etc.) correspondingly more.

Foods also contain ATP, derived partly from the original tissue (intrinsic ATP) and partly from microbiological contaminants. The proportion of intrinsic ATP finding its way into the cuvette, which therefore determines the lower limit of sensitivity of the method to microorganisms, obviously depends on the efficiency of the extraction method in separating the microbiological ATP. Using the methods recommended by the manufacturer, the Luminescence Biometer yields typical values for foods shown in Table 1. A fuller evaluation is given by Sharpe, Woodrow and Jackson (1970).

During incubation, the intrinsic ATP content of the food may fall, whilst bacterial ATP increases. Initially, that is for foods in a fit state to leave a factory or be consumed, the intrinsic ATP level can be seen to be

TABLE 1. ATP content of common foods

Sample	ATP (fg*)/g
Ice cream	$1 \cdot 2 \ 10^7$
Plaice fillet (frozen)	$9 \cdot 0 \ 10^6$
Beef	$5 \cdot 6 \ 10^8$
Bacon (prepacked)	$4 \cdot 5 \ 10^6$
Peas (frozen)	$5 \cdot 3 \ 10^7$
Milk	$3 \cdot 2 \ 10^7$
Yoghurt	$5 \cdot 4 \ 10^9$
Beef/mushroom soupmix	$1 \cdot 0 \ 10^{10}$
Soft margarine	$2 \cdot 2 \ 10^7$

* fg, femtogram.

so high that determination of bacteria by the recommended methods is not feasible. At some point in the life of the food the intrinsic ATP may decay, and bacterial ATP increase sufficiently, that determination of bacteria by this method is feasible. An improved situation can be obtained by incubating the sample with apyrases, which destroy c. 90% of the intrinsic ATP. Further improvement may be also obtained by studying the mechanics of removal of bacteria from a particular substrate, in order to optimize the ratio of bacteria to debris extracted from the food.

General comments

The Luminescence Biometer will detect approximately 500 fg ATP, which is equivalent to c. 1000 bacteria cells. This is injected as a $0 \cdot 01$ ml aliquot. Since most workers would need to handle 1 ml quantities, the sensitivity of the technique is best quoted as c. 10^5 cells/ml.

The short elapsed time (less than 10 min) before results are available could be valuable in some situations. Total technician time required when using the Biometer is about the same as for standard bacteriological counting methods. Reagents are relatively expensive, but the reduced cost of labour for washing and sterilizing probably compensates for this. It is doubtful whether reagent preparation for the Biometer could be entrusted to unskilled staff.

The Luminescence Biometer itself is simple to handle, easy to get used to and rapidly inspires a sense of confidence in its performance on suitable samples. The butanol, and particularly octanol, are rather objectionable however. However clean one tries to be, octanol can quickly taint glassware, washbasins, wastebins and fingers, from which it can be removed only by frequent rinsing with alcohol or acetone. The smell is unpleasant even in low concentrations and can cause headaches. If the extraction process was automated, most of this unpleasantness could presumably be eliminated. A completely automated instrument has recently been described in which the reagents are dispensed sequentially into sample vials inside a light-tight housing. The instrument processes 1 sample/min and samples are analysed in 15 min.

The usefulness of ATP determination as an indicator of bacterial contamination is, as has been seen, limited by the background of intrinsic ATP in foods. Williams (1971) takes a rather pessimistic view of its applicability to foods. For pure culture work, or for determinations on say, cooling water or other easily filterable liquids, these limits do not apply and, provided that a sufficiently large volume can be filtered, there is no reason why the ATP method could not be used down to quite low (say 10 organisms/ml) concentrations of bacteria. There is one aspect of ATP determination which makes the method potentially superior to conventional quantitative bacteriological techniques, and this is that it measures *biomass*, rather than the numbers of reproductive or colony-forming units. The total cellular mass of, for example, a long and strongly linked *Streptococcus* chain is a better measure of the contamination level represented by this organism than the single colony produced by such a chain on an agar plate. The method could also be potentially superior for the determination of contamination by moulds for which the existing quantitative methods are notoriously inadequate. Realization of the full potential of the method for foods must wait, however, until satisfactory methods of separating the bacteria from their substrate have been worked out.

The Methodological Approach

Few bacteriologists disagree that fresh looks at the methodology of bacteriology can result in improved analysis times, reduced materials consumption or increased output. There is no shortage of original ideas for streamlining bacteriological testing. The main problem is to see the usefulness of innovations in the perspective of the complete test, and it frequently happens that what may at first seem a good time-saving idea is of insignificant advantage when it is actually applied.

An essential preliminary to methodology modification, therefore, is a work study taken over the whole of a technique, from say, receiving the powdered raw media to the disposal of apparatus in an incinerator. Only when this has been done can the portions of the technique worth attacking be really seen. Such a breakdown of labour and costs associated with the analytical bacteriology (total viable aerobic count by pour plate) is shown in Tables 2 and 3. This type of study is even more important in the automation approach discussed later. From Table 2 it can be seen that the

TABLE 2. Breakdown of technical time in pour plate count*

	%
Attention to agar steaming/cooling	3
Collection of Petri dishes, bottles, scissors, etc.	12
Accurately weighing sample	24
Homogenization	13
Opening packs and labelling Petri dishes	14
Diluting, inoculating dishes and adding agar	22
Inspecting and counting plates	12
TOTAL	100

* Generally 5 or 6 dilutions, duplicate plates from each dilution. All media etc. taken from distributed stock.

TABLE 3. Breakdown of costs in pour plate count*

	%
Direct	
Media	3·0
Peptone	0·3
Distilled water	1·0
Petri dishes	9·6
Replacement pipettes	7·5
Replacement caps	1·4
	(22·8)
Indirect	
Depreciation of all items (homogenizers, autoclave, incubators, etc.)	7·4
Support labour	
Making ready homogenizers	7·8
Making ready pipettes	15·0
Filling and sterilizing bottles	11·0
	(33·8)
Technical labour	36·0
TOTAL	100·0

* Services (gas, electricity, etc.), management, maintenance costs, etc., not included.

technical operations associated with sample weighing and homogenizing contribute a large fraction of the time spent in counting. But reference to Table 3 shows that considerable additional time is spent in washing and re-sterilizing homogenizer cups. Thus, the fundamental problem in this area is really one of the homogenizers, and this whole aspect, rather than just sample preparation was attacked when the Stomacher homogenizer was developed. Similarly, in developing the Droplet Technique, a streamlining of not just one area of counting, but a reduction or elimination of un-necessary technical stages and service work was aimed at. The effects of modifying the methodology can be very significant, but at the same time, however good it is, any innovation is doomed to failure if it does not have a psychological impact making its introduction pleasant to the technicians who must use it.

Stomachers and Stomaching

This entirely new mixing technique was developed in order to eliminate the conventional hand-held or bench-top homogenizer, which requires considerable labour to re-sterilize its cup or probe before it can be re-used. Conventional homogenizers often have other disadvantages; for example, high capital outlay for adequate numbers of cups or probes, high noise level, undesirable temperature rises during sampling, and high maintenance costs. Alternative methods of sampling such as vortex stirring, ultrasound, scraping, spraying, and electrophoresis have been proposed but not widely accepted.

It is axiomatic that, during mixing, some or all of the mixer surfaces contact the specimen and become contaminated. If the labour needed for re-sterilizing these surfaces is to be avoided, then either the mixer must rapidly and automatically carry out a sterilizing routine, using for example, bactericidal solutions, or else all of the surfaces contacting the sample must be made disposable. This last approach was used in the Stomacher, and the generic name has been given to a family of devices which contain the sample in the cheapest of all possible containers—a plastic bag—and apply mixing forces to the sample through the flexible walls of the bag. For bacteriological purposes, satisfactory bags can be obtained with guaranteed sterility. Stomachers of various sizes are now available from A. J. Seward & Co. Ltd (PO Box 1, 6 Stamford St., London SE1).

All of the devices feature a means of temporarily sealing the sample and diluent inside the bag and of applying forces to the outside of the bag by means of paddles, wheels or rollers. The Stomacher shown in Figs 3 and 4 is a simple and successful design, particularly suitable for microbiology, but a range of devices handling up to or over 3 kg has been investigated.

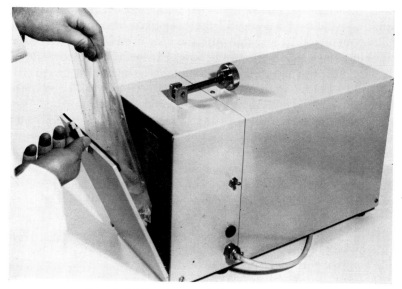

FIG. 3. Adding the sample bag to a Stomacher.

FIG. 4. With the Stomacher door closed the bag is trapped and temporarily sealed.

The bag is sealed near the top by being trapped between a rubber pad on the door and the bevelled edge of the case as the door is tightened. The large knob on top of the machine pulls the door firmly into contact with the bevel. When the machine is switched on, two paddles side by side alternately pound the bag and compress its contents against the door. A constant speed motor (230 rev/min) provides power to the paddles via two eccentrics, and resilience is built into the system by the use of rubber connecting rods. Excessive pressure build up, or stalling when incompressible samples are encountered is thus avoided. After samples for counting have been taken, the bag and its contents are thrown away. The Stomacher is then immediately ready for re-use and no cross-contamination can occur.

Removal of bacteria is brought about partly by violent shearing forces as the liquid is swept from side to side, and partly by the series of rapid compressions the sample experiences as it comes under the paddles. This repeated "sponging" action probably removes even deep-seated bacteria, which may be present in, for example, veins and capillaries in meat if the sample does not disperse completely.

Using the Stomacher and the recovery of bacteria to be expected

The sample is weighed into a 7×12 in. bag (e.g. Type BA 6031, A. J. Seward & Company Ltd., London, for the model shown) and diluent added. The bag is then placed in the processing compartment of the Stomacher, and the door closed and tightened before the machine is switched on. For soft samples, e.g. boiled fruits or comminuted meats, complete sample dispersal occurs almost instantaneously (5 sec). About 30 sec is recommended for routine work. It is not necessary for the sample to be disintegrated into a slurry for the bacteria to be removed, and with relatively tough materials, such as runner beans, complete sample disintegration should not be expected.

The recovery of bacteria from common foods, swabs and fabrics, by stomaching compares excellently with similar samples run on MSE-"Ato-Mix" blenders (Table 4). With meat samples containing about 95% fat the recovery by stomaching is $c.$ 50% after only 30 sec, but increases with time so that 60 or 90 sec gives a comparable yield. The sample is easily visible through the walls of the bag, and can be reprocessed if its appearance is unsatisfactory. No reduction in viability has been found after prolonged (e.g. 5 min) stomaching, so that processing time is not critical.

A variety of flexible bags can be used, for example, polyethylene, polyester, rubber and laminates. A standard thin-walled polyethylene bag, preferably base-welded, is adequate however, and is generally to be preferred on the grounds of cost, ease of opening, and freedom from toxic effects. Flexibility is far more important than "toughness", and bags of stronger appearance may give disappointing results. Sharp objects such as bone splinters may make pinholes through which the suspension will leak, and samples containing such objects are best avoided. Generally, leaks occur far less frequently than with conventional bottom drive blenders. The wooden sticks of swabs do not damage the bags.

TABLE 4. Products for which comparisons of bacterial recovery by Stomacher* and "Ato-Mix" have been made

	Comparison
Beef cuts	NS
Beef cuts (50 % fat)	NS
Beef cuts (95 % fat)	S < A
Beef cuts (15 sec stomaching)	NS
Chicken	NS
Comminuted meats (sausage, burgers)	NS
Reformed beef (15 sec stomaching)	NS
Pastries	NS
Fruits (apple, peach)	NS
Vegetables (carrots, cabbage, turnip, peas)	NS
Vegetables, dried (mushroom, carrot, onion, cabbage, swede)	NS
Reformed potato	NS
Fish (kippers, cod, prawns)	NS
Fabrics	NS
Swabs	NS

* 30 sec stomaching unless otherwise indicated; NS no significant difference (P = 0·05); S < A stomach counts lower than "Ato-mix" counts. Total comparisons 301.

Advantages of stomaching for bacteriological sample preparation

Stomaching has many advantages as a sample preparation method for any bacteriology laboratory dealing with reasonably soft materials, particularly in regard to elimination of labour for recycling cups or probes. Capital outlay is small and running costs negligible. Bags require very little storage space, and can be obtained with guaranteed sterility. These bags are excellent for taking and transporting samples from the factory. Sample preparation time is usually short.

Temperature rise during stomaching is negligible (0·8°/min from ambient). This contrasts markedly with some types of blender, for which temperature rises up to 20°/min can be recorded, and which can pasteurize samples if left on too long. Noise level is low; the Stomacher makes a dull squelch, which is far less irritating than the whine of bladed homogenizers.

Samples can be processed whilst still deep frozen. The resilience of the

Stomacher paddle system allows the machine to ride over the temporary obstruction posed, for example, by a piece of frozen meat, and the stomaching action ensures very rapid thawing as diluent is repeatedly driven deep into the sample structure. The rapid thawing produces an unmistakable change in the note of the Stomacher.

The debris level is frequently much lower than that obtained from conventional homogenizers. This can make colonies on Petri dishes very much easier to spot (Fig. 5). Several samples can be processed simultaneously. The Stomacher shown in the Fig. 3 will handle at least 400 ml. Thus, if 100 ml of sample and diluent are used, up to four bags of samples can be processed together, thereby effecting even greater economies in time. Alternatively, 2 bags containing 200 ml can be processed. The disruptive action is sometimes reduced when several bags are processed, so that it is wise to check on the length of processing time required for each type of sample before routinely doing this.

The mixing action continues with very viscous samples, and even with semi-solids or powders. Thus the ratio of sample to diluent is not important, and high ratios can be satisfactorily employed when low contamination levels are expected. Moreover, the Stomacher will mix many other materials such as batters, glues, and feed additives.

The whole device is light, and easily portable with a supply of bags. A Stomacher, combined with the Droplet Technique (discussed below), forms an excellent portable testing method.

The instrument is safe, both from the likelihood of causing serious injury or the production of dangerous aerosols. The resilience of the paddle system ensures that a salutory but mild bruising is the worst result of putting fingers inside whilst the machine is running. Since the bag is sealed no aerosol can escape. Additional protection against the possibility of leakage of pathogens from pinholes may be obtained by placing the sample bag inside another of different material. The design ensures that, in the unlikely event of catastrophic bag failure, the motor and electrical fittings are completely protected. The paddle compartment can be swilled down and wiped with tissues.

The Droplet Technique and Apparatus for Counting Bacteria

Within the limits defined by "agar methods" for counting bacteria, there are a group of techniques related to one another by their dependence on the basic processing system:

(a) suspending the bacteria in water (or other diluent);

(b) making a range of dilutions;

(c) planting the bacteria in or on a nutritive agar;

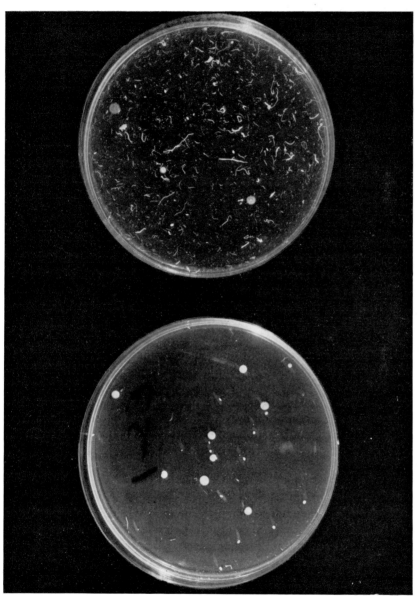

FIG. 5. Corresponding plates prepared from a meat product by a Stomacher (left) and a conventional homogenizer (right). Reduced debris level makes the plate from the Stomacher very much easier to read.

(d) waiting whilst they multiply;

(e) counting the colonies, and

(f) calculating the original concentration level.

The techniques in the group differ from one another only in the manner in which the basic processes are carried out, or omitted. The standard pour and spread plates, the agar sausage, the Miles and Misra, and the Frost microscopic methods, for example, all differ greatly in the way of making (or not making) dilutions, inoculating the agar, the means of holding the agar, incubation time, method of counting, and of relating the count to the sample, but each was developed to apply the same basic processing system to best advantage under a particular set of conditions.

The Droplet Technique developed at Colworth is a new method, (Sharpe and Kilsby, 1971) which is likely to be valuable particularly in factory and research laboratories where many samples must be screened. Like all the other methods noted above, it is an attempt to apply the basic system to best advantages in a particular environment, and like these it has both advantages and disadvantages. The instructions given should not be interpreted as a doctrine but only as a set of principles which can be "bent" to suit particular needs. For example, when 3 centimal dilutions are suggested, there is no reason why 2 decimal dilutions should not be made if it seems more appropriate.

When compared with a standard plate count, the Droplet Technique allows bacteriological counts to be made cheaply and quickly. Three to five times as many samples can be processed per day and overnight incubation is sufficient in most cases. Counts made using this method correlate well with those made using the standard pour plate method. The technique saves considerably on materials and can be done with standard laboratory equipment. Two specialized pieces of apparatus were developed at the same time as the technique, however, and the use of these allows the advantages of the technique (simplicity, rapidity, economy) to be realized fully. These aids are fully described below.

General principles of the Droplet Technique

The technique is a miniature pour plate method. Unlike the pour plate method, in which one or more Petri dishes are used for each dilution, the Droplet Technique requires only one half or one quarter of a dish for a sample, each dilution being plated out as a series of droplets (0·1 ml). The bacteria are planted in the agar. Each droplet of agar can then be regarded as equivalent to that in a Petri dish, and it contains miniature colonies after incubation.

This approach reduces materials consumption and incubation times without incurring the manipulative difficulties found in true micromethods. At the same time, it retains the inherent simplicity of a pour plate method and avoids the time consumed in drying agar plates required in, for example, the Miles and Misra method, which is the spread plate equivalent of the Droplet Technique.

Dilutions are made in the molten agar. This unusual procedure is quite practical, and causes less thermal stress in bacteria than the ordinary pour plate technique. Pipettes are a matter of choice, cheap disposable Pasteur pipettes being generally preferable if the technique is done entirely manually. If the Diluter/Dispenser is used, either Pasteur pipettes or even cheaper polypropylene drinking straws can be used. The Diluter/Dispenser allows 1 in 10 (decimal) or 1 in 100 (centimal) dilutions to be made with equal ease.

Droplets are dispensed either as three drops from Pasteur (dropping) pipettes or with the Diluter/Dispenser. Five droplets per dilution are normally dispensed, but this is a matter of choice. After incubation colonies are counted under low magnification, say $\times 10$. A hand lens or plate microscope can be used, but the viewer (Fig. 9) is to be preferred. In counting there is frequently a choice, depending on how the dilutions have been made up. For example, five droplets might be counted, each containing say 10 colonies, or one droplet containing 100 colonies.

Very low bacterial concentrations are not countable. For an average of one colony per droplet in the first dilution, counts in the sample will be c. 100–1000/g depending on how the sample suspension was made. The Droplet Technique is, therefore, suitable for samples containing say 1000 bacteria/g and upwards, and, unless pure cultures are being used, should be regarded as a total viable count method.

Droplet preparation

The medium is prepared in 9 ml amounts in 1 oz screw cap bottles and sterilized by autoclaving. For each total viable count, 2 or 3 bottles of medium should be melted and cooled to 45° in a water-bath.

Label a 3 in. unpipped plastic Petri dish with the sample number; dilution numbers are not needed. Draw or imagine three parallel lines across the outside of the dish. Insert a sterile dropping pipette or straw into the holder of the Diluter/Dispenser.

Pipette 1·0 ml of the sample suspension into the first of the bottles of cooled medium. Mix the contents of the bottle by shaking it gently or using a "Whirlimixer" (Fisons Ltd, Loughborough, Leics.), and dispense five 0·1 ml droplets in the Petri dish along the first of the marked parallel lines. Allow another 0·1 ml to fall into the second bottle of cooled medium.

Mix and, using a fresh sterile pipette, dispense droplets from the second bottle in exactly the same way, placing five droplets along the second of the marked lines on the Petri dish (Fig. 6). These droplets contain 1/100 the concentration of sample compared with the first row. If necessary, make a third 1/100 dilution in the remaining bottle of medium and place the droplets along the third line of the Petri dish.

The Petri dish thus contains one total viable count in 3 rows of 5 droplets. With a little practice all the droplets for one count can be formed in about 40 sec. A second set of droplets for another count can then be placed in the lid of the dish. For products with less than 10^7 bacteria/g, only 2 dilutions need to be made and four samples can be plated out in one Petri dish.

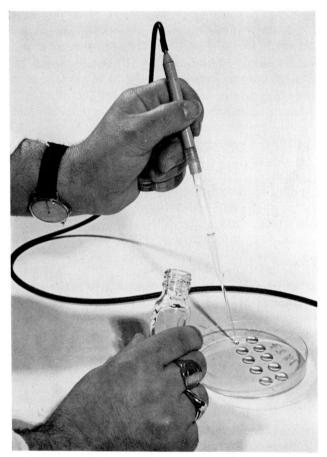

FIG. 6. Dispensing droplets (0·1 ml) with the aid of the Diluter/Dispenser.

Incubation

The droplets should be incubated at the temperature normally used for pour plates, but a check should be made on the length of time required. As a general rule, overnight incubation (20–24 h) will be sufficient to bring all colonies to a countable size (Fig. 7) if 48 h would be sufficient for pour plates. Exceptions may arise if the bacteria show a considerable lag phase as, for example, with the psycrophilic bacteria often found on fresh or frozen fish. In such a case, 48 h incubation may be required.

Droplet counting

Colonies are counted at about ×10 magnification, using a lens, plate microscope or the viewer (Fig. 9). Take a mean count for up to 5 droplets at the chosen dilution. At least 20–30 colonies should be counted. For example, if there are 5 colonies/droplet, count all five droplets. If the count is greater than 100/droplet, only one or two droplets should be counted. Up to 200 colonies/droplet can be counted conveniently. No difficulty will be experienced in identifying which dilution pertains to each row of droplets.

Calculating

The factor to be applied obviously depends on the dilutions made, but it must be remembered that droplets contain only 0·1 ml of suspension so that the count *per ml of that dilution* is ×10 the droplet count, i.e. the droplet itself represents one decimal dilution. For example, suppose there were 25 colonies per droplet in the second row, and that the dilution sequence was:

	Factor
Sample to liquid in homogenizer (10 g to 90 ml)	×10
Suspension to bottle No. 1 (1 ml to 9 ml)	×10
Bottle No. 1 to bottle No. 2 (0·1 ml to 9 ml)	×100
Bottle No. 2 to 0·1 ml droplets	×10

The count is therefore $25 \times 10 \times 10 \times 100 \times 10$ or $2\cdot5 \times 10^6$/ml in the sample.

In describing the technique, dilution factors have been rounded to factors of 10. Thus, although 1·0 ml transferred to 9·0 ml gives a 1:10 dilution, 0·1 ml transferred gives, strictly, a 1:91 dilution—*not* 1:100. The error is unlikely to be noticed. If dilutions are always made to the same factor, i.e. either ×10 or ×100, the volumes of media in the bottles may be adjusted to eliminate this error entirely.

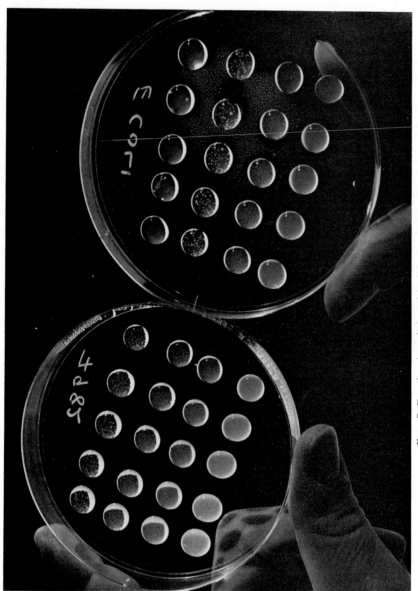

FIG. 7. Droplets in a dish and its lid after 24 h incubation.

The Diluter/Dispenser

The device consists of a foot control connected to a handpiece by a length of flexible tubing (Fig. 8). When the pedals are operated, liquid can be taken into or expressed from a pipette or straw held in the handpiece. No liquid enters the flexible tubing. One pedal (A) will suck and eject 1·5 ml in the pipette. This is used for mixing and filling the pipette. The other pedal (B) delivers 0·1 ml each time it is depressed and released. There is no need for judgement. By sucking up 1·5 ml, and dispensing 5 × 0·1 ml droplets, 1·0 ml is left in the pipette. Then either 0·1 ml can be transferred to the next bottle of agar, to make a 1:100 dilution, or the whole 1·0 ml to make a 1:10 dilution.

Repeated operation of pedal A with a pipette immersed in molten agar will cause the pipette to take up more than 1·5 ml. This is unavoidable and

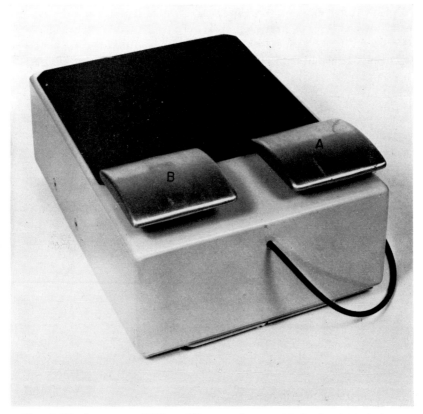

FIG. 8. The foot operated Diluter/Dispenser for use with the droplet technique.

is caused by expansion of the cold pump air as it passes through the warmed pipette barrel on the downstroke. To avoid introducing an error when 1 in 10 dilutions are made by transferring the 1·0 ml of agar remaining after dispensing 5 × 0·1 ml droplets, it is advisable to mix using the pipette from the previous bottle and to take up 1·5 ml in the new pipette with a single stroke. The pump operated by pedal A can be adjusted to suck up and deliver 0·5–1·7 ml. Both the Dilutor/Dispenser and viewer are available from A. J. Seward & Co. Ltd (PO Box 1, 6 Stamford St., London, SE1).

The Magnifier/Viewer

The Viewer is a "condenserless" projector, with a ground glass screen directly facing the observer (Fig. 9). To use it, an open Petri dish containing the droplet under examination is placed over the illumination hole. A 50 W, 12 V tungsten/halogen lamp situated immediately beneath this hole throws out a divergent beam which is refracted towards the lens by the droplet itself. The droplet is its own condenser and no focusing is needed. The beam is reflected by a prism and then by a mirror on to the screen.

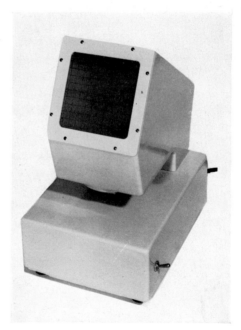

FIG. 9. The Viewer designed for use with the droplet technique.

Droplets and their contents are magnified $\times 10$, appearing on the screen at about the size of a standard 90 mm Petri dish (Fig. 9). Bacterial colonies generally have characteristic ellipsoidal shapes and are straw to dark brown in colour. The colony images may be touched with a felt tipped pen during counting, and the screen subsequently wiped quickly with a damp tissue. Alternatively, the grid lines on the screen may be used to prevent over counting. It is generally easier to count several droplets of low colony density than one of high density.

The condenserless system is provided specifically to give a very bright image. No difficulty should be found in seeing the image in strong sunlight. As the lamp is relatively powerful and very close to the Petri dish, the latter should not be left over the illumination hole any longer than is necessary to count the colonies otherwise the plastic may be distorted.

Performance of the droplet technique

As can be seen from Table 5, the technique generally gives results for total viable aerobic counts which do not differ significantly from those obtained by standard pour plate methods. Occasionally, with bacteria which are particularly aerobic, the droplet technique gives greater recoveries than the pour plate, as a result of the greater surface/volume ratio and consequently better aeration of the droplets. Evaluation of colony numbers is occasionally difficult, as for example, in droplets made from bacon or mutton, because of the high proportion of fatty debris, but in most other samples colonies are easily distinguished from debris, even in the highest concentrations.

The technique is learned very easily when the Diluter/Dispenser and Viewer are used, and technicians quickly prefer its simplicity. There is a tendency not to regard it as a miniature method, and the appearance of

TABLE 5. Analyses made with Droplet Technique

Raw beef	$D \geqslant P$	Raw ham	NSD
Dried beef	NSD	Prawns (AFD)	NSD
Kidney	NSD	Cod (frozen)	NSD
Brisket	NSD	Peas	NSD
Mutton	NSD	Disinfectant lethality	NSD
Rolled Mutton	NSD	Skin bacteria	NSD
Chicken	NSD	Cooling water	NSD
Dried chicken	NSD	*Pseudomonas putida*	NSD
Beefburger	NSD	*Pseudomonas fluorescens*	$D \geqslant P$
Sausage meats	NSD	*Staphylococcus aureus* (spread plates)	NSD

Comparisons with poured plates. Total No. of comparisons 525; level of significance 0·05.

D, droplet count; P, pour plate count; NSD, no significant difference between plate and droplet counts, and $D \geqslant P$, sometimes greater, varies from sample to sample.

magnified droplets on the Viewer screen at approximately the size of a
normal Petri dish provides an excellent psychological stimulus to its
acceptance. Equally important is the ease with which the technique inte-
grates with, and even facilitates other types of analyses which may still
be done by standard methods. Diluter/Dispensers are used, for example, to
inoculate bottles for Most Probable Number (MPN) counts of coliform
organisms, to inoculate the surfaces of plates for *Staph. aureus* determination, and for dispensing broth or sera for confirmatory tests. The Diluter/
Dispenser virtually eliminates fatigue in diluting or plating, and the Viewer
fatigue in counting, particularly in comparison to methods requiring the
use of a microscope.

Significant savings are made on labour and materials costs. Labour
savings result from the reduced number of liquid transfers made during
each analysis, reduced manipulation of Petri dishes, elimination of pipetting and much writing on dishes, and the reduced volume of all apparatus
at the bench which necessitates less reaching and fewer trips to stores.
Diluting and plating on most samples requires approximately 40 sec and,
if only the droplet technique is being used, output can be increased threefold over conventional plating methods. At the same time, reduced consumption of media, diluent bottles and pipettes reduce preparative work.
Cost savings result from reduced consumption of Petri dishes (one quarter,
instead of 4 or more per count), pipettes (if disposable graduated ones
were used), and to a lesser extent from the smaller quantities of media
used.

Only a small amount of work has been done on the applicability of the
droplet technique to other types of bacterial count. There is no reason to
expect, however, that a variety of selective agars or atmospheres could not
be used, provided that the only criterion by which colonies are recognized
is their presence (i.e. not colour, zones of clearing, etc.). The technique
could be valuable in, for example, thermal inactivation studies on pathogenic anaerobic bacteria or spores, where great savings in anaerobic jar
space are possible.

A number of unusual features about the droplet technique occasionally
cause concern when it is met for the first time. One of these is its accuracy,
for which the data in Table 5 should provide immediate reassurance. Other
questions which have been raised from time to time are answered below.
Some can be answered factually, others are more subjective.

Ease of recognition of colonies

For most samples, the visibility of colonies in droplets is better at 24 than
on pour plates at 48 h, owing to their relatively greater size after magnifi-

cation. Colonies normally have a straw to dark brown, ellipsoidal appearance which is easy to spot against any ragged debris. Only with *Clostridium* spp, so far, have colonies been sufficiently ragged to be likely to be mistaken for debris. These, of course, are not met with in aerobic work. The difficulty of counting when a high proportion of fat is present has been mentioned, but fat droplets frequently appear sufficiently characteristic to not be falsely taken for bacterial colonies.

Effect of diluting in molten agar

Fears about the effect of thermal stress on organisms as a result of diluting in molten agar are unfounded. Thermal stress is, in fact, lower in the droplet technique than in the pour plate, since the *maximum* time for which any of the organisms are held at 45° is *c.* 40 sec. After delivery, the droplets cool to less than 30° in 10 sec. In contrast, a set of pour plates which, after pouring, may be held in a stack of 8–10 dishes, may require up to 10 min before the temperature of the middle dishes drops below 40°.

Colony crowding with respect to pour plates

The existence of say 200 colonies in a 0·1 ml droplet may suggest that the colonies are overcrowded and that poor growth will ensue. In fact, such colonies are relatively less crowded than on a pour plate, this rather surprising fact following from the Cube Law. If it is assumed that colonies in droplet and plate are equivalent in recognizability when their diameters subtend equal angles to the eye then, other things being equal (density of organisms in colonies, media concentration, etc.), the numbers of just visible spherical colonies which can be supported by droplet and plate respectively are given by:

$$\frac{\text{Number in droplet}}{\text{Number in plate}} = \frac{\text{Volume of droplet}}{\text{Volume of plate}} \times \frac{(Md)^3}{(Mp)}$$

where Md, magnification used to view droplet = 10, and Mp, magnification used to view plate = 1

For droplets of volume 0·1 ml, and plates of volume 15 ml this equation gives:

$$\frac{\text{Number in droplet}}{\text{Number in plate}} = \frac{0\cdot1}{15} \times \frac{(10)^3}{(1)}$$
$$= 6\cdot6$$

i.e. the droplet could, if necessary, support 6·6 times as many just resolvable colonies as the plate.

Other factors, such as the ellipsoidal rather than spherical shape of colonies have not, of course, been considered. However, in practice neither plates nor droplets would be counted with so many small colonies as to exhaust their nutritive powers and it can safely be assumed that the droplets contain at least as great a reserve of nutrients, if not greater, than a plate, and that overcrowding is unlikely to be a problem with droplets.

Ease of measuring out 0·1 ml droplets

When the droplet technique is carried out entirely manually—that is, when 0·1 ml droplets are dispensed from a dropping pipette—the usual considerations of accuracy apply. Technicians should follow the standard practice of, for example, holding the pipette vertically, releasing drops smoothly, and checking the accuracy of each box of pipettes. If the Diluter/ Dispenser is used, much less attention is required, since metering is done mechanically. The accuracy of manufacture of pipettes becomes irrelevant, and it is enough just to make sure that the tip of the pipette or straw is momentarily in contact with a dispensed droplet. The design of the dispenser, incidentally, virtually precludes the possibility of dispensing an incorrect amount, since unless the pedal is depressed completely the 0·1 ml pump will not deliver.

Eyestrain

A complaint frequently brought against miniature techniques is that the need to count colonies with the aid of a lens or microscope leads to eyestrain. Such a criticism can also be levelled against the droplet technique if it is used without the specially designed Viewer, although the low magnification required would normally allow a large, easy viewing bench-lens to be used. The Viewer used with the technique, however, gives a very bright screened image of droplets, directly facing the technician as he sits at the bench. With droplets magnified to the size of a normal 9 cm Petri dish, eye strain is no more noticeable than when counting plates. An electric counter can be used with the Viewer if required.

Automation Approach

This approach supplements the methodological approach by providing for the automation of all or a proportion of a conventional counting technique, with little or no attempt to alter the basic methodology. Its chief advantage is that, whilst potentially capable of yielding considerable increases in technical output, the instruments developed should be much more readily acceptable in the early stages to bacteriologists (who are justifiably conservative) than more radical approaches introducing techniques of less proven reliability.

In considering this approach, an additional breakdown of processes, beyond that illustrated in Tables 2 and 3, into "unit operations" is invaluable, and can lead to surprising conclusions about the merits of automation under particular circumstances. For example, the act of diluting can be broken down into individual commands such as: "*pick up pipette*; *pick up bottle*; *insert pipette into bottle*", and so on. In this way the complete process of counting bacteria by pour plates can be broken down into well over a thousand unit operations. Of these, less than 10% are actually concerned with the basic operations of diluting, plating and counting. For example, in diluting, which consists basically of transferring liquid from A to liquid B, the remaining operations are "wasted effort" in terms of the actual count, but are necessary to fit the technique to human hands. Because of this, introducing a machine to carry out one of the basic operations is unlikely to yield a significant time saving if the operator is still required to handle bottles, replaceable "pipettes", Petri dishes, etc. before and after the operation. Significant time savings are only likely to be achieved when the introduction of a machine completely eliminates the need for the presence of the technician at one or more stages—the more the better—in the process.

A pour plate preparing machine

In designing a pour plate preparing machine, the large fraction of total cost accounted for by and the central position occupied by pipetting were recognized. It was considered essential to carry out this step entirely automatically and aseptically yet at low cost. Two areas of pour plate counting were not considered for automation at this stage. These were:

(a) the sample preparation step, which has since been considerably facilitated by the development of Stomacher homogenizers, and

(b) colony counting, for which the cost benefit of any practical electro-optic device was likely to be marginal.

The relatively simple mechanical system employed in the pour plate machine saves or eliminates $c.\ 70\%$ of the total labour involved in counting and it was not reasonable to consider including these other stages of the process at present. All of the bacteriological processes carried out by the machine are integrated and, for most of the time that it is running, the operator need only put samples in one side of the machine and remove completed plates from the other side. By eliminating the technician from several successive stages the most efficient utilization of automation is achieved. In addition to saving technical time, however, elimination of the requirement for pipettes and diluent bottles provides a dramatic and equally significant saving in preparative time.

An experimental plate preparing machine is shown in Fig. 10. From a sample suspension it automatically prepares a stack of inoculated, agar-filled Petri dishes (labelled with sample and dilution number) ready for incubation. According to the setting of the controls it will select and prepare

FIG. 10. The automatic diluting and pour plate preparing machine seen from the sample-input side.

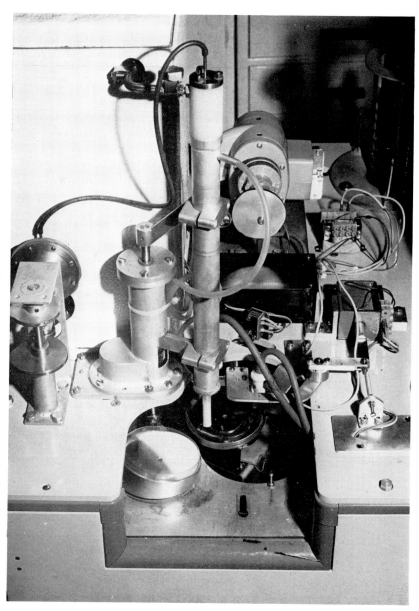

FIG. 11. An agar pipette transferring liquid from the sample pot.

up to 8 dilutions in each of 3 media (e.g. Violet Red Bile agar (VRBA) for coliform counts; Plate Count agar (PCA) for total aerobic counts, and Malt Extract agar (MEA) for yeast and mould counts). The various operations are carried out sequentially and simultaneously. All materials are handled in bulk, and fresh, sterile "pipettes" for dilutions are produced within the machine as required.

Dilutions are made within the Petri dish itself, the residual liquid, after transferring an aliquot, serving as the inoculum. In its present form the machine adds 10 ml diluent to each dish and transfers 1·1 ml aliquots through the dilution range, thus achieving a set of decimally diluted inocula, each of 10 ml. To these, 7 ml of two and a half strength agar is eventually added. The final content of each Petri dish is, therefore, 17 ml of normal strength agar. Suitable concentrated nutrient solutions are used as the diluent, and agar is added as a pure solution. This avoids keeping nutrients for long periods at elevated temperatures; the agar itself is not significantly affected by so holding.

Pipettes are produced automatically and as required from the concentrated agar used to fill the dishes. They are extruded as cylinders from a water-cooled die (Fig. 11) and discarded after use. Cutting is done with a stainless steel blade, sterilized by being heated electrically to c. 250°. A constant supply of sterile pipettes is thus assured and carry over of contamination eliminated. A separate diaphragm pump, connected through the centre of the die, controls the volume of liquid taken up and ejected by each pipette. The shape of the agar cylinder used to transfer liquid is therefore unimportant provided that it is long enough to make contact with the diluent in the Petri dish. The present machine uses the clover leaf form (Fig. 12) to obtain a large cross-sectional area without tendency to drip.

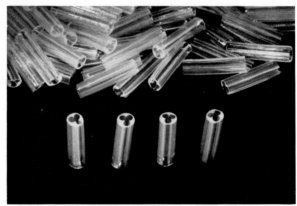

FIG. 12. Used agar "pipettes".

In its present form the machine prepares one filled and labelled dish every 15 sec, or 2000/8 h day. Its speed can be increased if necessary. The number of samples which can be done is determined by the number of dilutions required. If a typical output requirement, of, say, 2 dilutions in VRBA, 6 in PCA and 2 in MEA are repeatedly called for, 150 samples can be processed easily in a normal working day. Whilst doing this, of course, the machine requires no pipettes, diluent bottles or agar bottles.

Layout

Bulk diluent stores and metering pumps are located in the lower part of the machine so as to avoid the need to lift heavy weights. Diluent bottles up to 10 l can be accommodated, allowing up to 4 h uninterrupted operation. The agar preparation unit, which was originally located inside the machine has been removed and now forms a separate module. Agar stores can be replenished whilst the machine is running.

All bacteriological operations (dispensing Petri dishes, removing their lids, performing dilutions, adding agar, mixing, replacing lids, and labelling) are carried out in the covered portion below the main working surface of the machine. Dishes are held in a rotating cross-shaped carrier which indexes round to allow the various operations to be carried out at different positions.

Petri dishes and prepared agar are stored above the main working surface. Dishes are added to the rotating store (capacity 200) whenever convenient. Agar (10 l) can be pumped into the storage tank as soon as it is ready, leaving the preparation tank free for rinsing or further preparation.

Preparation and handling of consumables

Diluents

These are prepared in bulk, away from the machine, in bottles fitted with supply tubing and delivery nozzles, which are sterilized by autoclaving at the same time as the diluent. A set of autoclavable piston pumps, designed so as to prevent entry of contaminants, deliver diluents to the Petri dishes (Fig. 13). The delivery characteristics of these pumps are greatly superior to those of peristaltic pumps which were earlier evaluated on the machine.

Diluents are prepared as solutions of the media nutrients, at about $\times 1 \cdot 7$ the intended final concentration. They are used at room temperature and cannot, therefore, deteriorate through being maintained at high temperatures.

Agar

For routine work, agar can be prepared in the electrically heated steam-jacketed make-up tank. Agar powder and cold water are added to the tank about 35 min before a solution is required. A centrifugal pump circulates the agar suspension during boil-up, and is later used to pump the solution to the storage tank where it is held at 60–80° until used. An alarm circuit in this tank gives warning of low level in time for a fresh mix to be made.

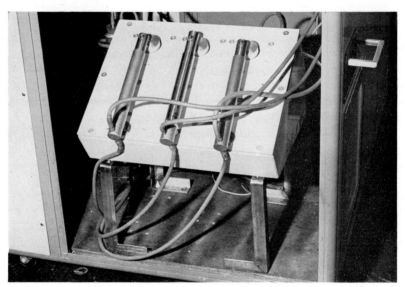

FIG. 13. Diluent delivery module. Each pump delivers 10·0 ml/stroke. Each diluent produces a different agar composition.

Petri dishes

The rotating holder can take up to 200 disposable plastic Petri dishes in 8 stacks. As one stack empties the holder indexes round to bring a fresh stack above the dispenser. Fresh dishes can be added at any time.

Electricity and water

The normal mains electric supply is required. The machine consumes *c.* 3·5 kVA with all units working, most of this being taken by the agar preparation unit. Cooling water is required for pipette extrusion, and for a condenser on the agar preparation unit steam-jacket. Coolant temperatures up to 30° can be tolerated, but "pipette" quality decreases rapidly above 25°.

Operation

Start-up operations are completed quickly. These include switching on services, and adding agar powder (conveniently by volume) and water to the preparation tank. Between this point and beginning work the following items are attended to:

(a) Renewal of diluent bottles, and pumps;

(b) filling the Petri dish store;

(c) removing a disinfectant cup from the base of the pipette dispenser (if the instrument has been left overnight);

(d) pumping boiled agar solution to the storage tank (if much work is to be done, the preparation tank is refilled immediately, but not boiled; otherwise, it is rinsed and emptied), and

(e) running one complete set of dilutions in order to fill and clear the various supply lines.

During this time the first sample suspension will have been made ready. Preparing a set of dishes involves no more than:

(a) Placing a pot containing the sample suspension in the machine;

(b) pressing buttons to programme the required number of dilutions for each agar, and

(c) pressing the "Start" button.

The machine then completes its bacteriological operations whilst the next sample is prepared. Before, or soon after inserting the next sample, the operator removes the stack of prepared dishes. Only one sample is accepted at a time.

Sequence of events from pressing "Start" button

All operations are carried out with the dishes seated in rings at the ends of a fairly complex cross-shaped carrier mechanism. The carrier indexes around one quarter of a revolution for each operation. As "Start" is initiated:

1. A Petri dish (dish 1) is dispensed (Station 1) and carried around to the dilution position (Station 2). On the way its lid is removed and the dish is brought into an inclined position. At Station 2 the dish engages with a rotating disc and itself begins to rotate.

2. An agar pipette is dispensed. The dispenser moves to the sample pot, withdrawing $1 \cdot 1$ ml, then to the first Petri dish where the pipette is emptied. Diluent (diluent 1)–$10 \cdot 0$ ml–is pumped into the dish. The contents of the dish mix rapidly as a result of its rotation. The pipette is cut and falls into a waste receptacle. A new pipette removes $1 \cdot 1$ ml of the decimal dilution. A fresh dish (dish 2) is dispensed at Station 1.

I

3. The carrier indexes around again, taking dish 1 to Station 3 (agar addition) where it again rotates whilst in an inclined position. Agar is pumped into the dish and mixed. At the same time the events (1) and (2) are repeated with dishes 2 and 3, except that the pipette does not move to the sample pot.

4. The carrier indexes again, taking dish 1 to Station 4. On the way its lid is replaced. At Station 4 the dish is pushed from beneath by the printing mechanism and rides over a set of catches to form the first dish of a stack. The events (1)–(3) are repeated.

5. At a point determined by the operator when programming the machine for the sample, the pipette dispenser again moves to the sample pot, picking up a fresh aliquot of the suspension. The sequence then continues using diluent 2 to obtain a second set of dilutions, and is finally repeated with diluent 3.

6. As the last dish is labelled and stacked the machine stops unless a new sample has been inserted. A push-button programming system allows the programme to be held for the next sample if no change is required, and in this case the operator can at any time instruct the machine to proceed directly with the new sample as soon as it has completed the final dilution of its first sample.

Speed and time saving

The machine will complete a normal dilution-range count preparation from three media in 2·5–3 min, and smaller count ranges in less than this. The operator can use this time to prepare the next sample. With many types of material, sample preparation is the rate determining factor, and two or more operators can use the machine simultaneously. The potential output exceeds the capabilities of most bacteriologists.

Labour saving on preparative work is very marked. All operations concerned with pipettes, and small bottles of agar or diluents (uncapping, washing, refilling, recapping, etc.) are eliminated. Agar preparation occupies very little time, equivalent to that spent steaming separate bottles for a normal manual count. Diluent (nutrient) preparation time is equivalent to that required to prepare the same volume prior to dispensing into bottles for the manual count technique. Large diluent bottles may be kept in the machine overnight for re-use the next morning, provided the nozzles are removed from their brackets and immersed in formalin. Total labour saving appears to be c. 70% over the whole count, with cost savings of around 47% on each count.

Accuracy

The machine prepares excellent decimal dilutions, well mixed with agar, showing no sign of "carry-over", and at a maximum temperature of 40–42° after mixing. It is reasonable, therefore, to expect the accuracy of the machine to be equal to or greater than that of a normal manual count. As an example, a set of 144 counts obtained as quadruplicated analyses on 36 specimens (meats, vegetables and swabs) gave the following results ($\log^{10}$ counts/g):

Machine: Mean count, 3·89; estimate of variance, 0·0027

Manual : Mean count, 3·87; estimate of variance, 0·082

indicating that the variability of counts made by the machine is significantly less ($P < 0.1\%$) than those done manually.

In the absence of nutrients, and at the high storage temperature, multiplication of bacteria in the boiled agar is most unlikely. However, as prepared by boiling, the agar has a very small but detectable level of residual bacterial contamination. In normal laboratory atmospheres, 1·5–6% of plates have one contaminant colony. Most of this, in fact, is due to aerial contamination, and for critical work such as sterility testing, addition of a filtered air supply means may be desirable. For normal quality control work with foods, even a 6% chance of finding one contaminant colony on a plate leads to an undetectably small error, since it is unusual for plates containing fewer than 20 colonies to be counted. It is quite feasible, of course, to prepare agar outside the machine and sterilize by autoclaving, but in most instances the extra labour is unjustified.

Only one plate is prepared from each dilution. Many bacteriologists, however, like to inoculate Petri dishes in duplicate from each dilution. The effect of such duplication on the accuracy of the count is questionable and it would be far better if the dilution sequence was also repeated; better still if the whole product was resampled. This is feasible with the machine described here. If the utmost in accuracy is required, therefore, it is more reasonable to take two samples from different parts of the product, rather than to duplicate inocula from particular dilutions. Since the accuracy of the machine is greater than most manual workers achieve, duplication of counts from one sample is hardly justifiable.

Future developments

Many aspects of the machine are being redesigned so as to improve accessibility, versatility, and general smoothness of operation. For example, increasing the number of media dispensed to at least four, and adding the ability to prepare uninoculated plates of a fifth medium for subsequent manual surface inoculation should allow the machine to satisfy any reasonable subsequent demands on performance. A range of modules to allow the machine to perform a wide variety of quality control or research tests is also being designed. At the same time, evaluation work is being done on other types of count to ensure that the pour plate approach embodied in the machine can be used to its full in quantitative bacteriology.

Note added in proof

Since this article was written considerable advances have been made in all Colworth-designed instruments. In particular, the Diluter-Dispenser has been developed into a very rapid electrically-operated unit fitting into the Viewer base, together with an electromagnetic Counter. This has produced a very compact droplet technique apparatus. The automatic pour plate machine has also been developed further along the lines indicated. All items are now marketed by A. J. Seward & Company Ltd.

Acknowledgements

The author gratefully acknowledges the assistance of D. R. Biggs, W. Crabb, E. J. Dyett, B. Fletcher, A. K. Jackson, D. C. Kilsby, V. W. Klicska, G. R. Lloyd, R. J. Oliver, F. J. Rupp, B. J. Szablowski, R. Tucker, M. N. Woodrow, F. Vout, and many others during the production or evaluation of these techniques and devices.

References

ADDANKI, S., SOTOS, J. F. & REARICK, P. D. (1966). Rapid determination of picomole quantities of ATP with a liquid scintillation counter. *Analyt. Biochem.*, **14,** 261.

McELROY, W. D. (1947). The energy source for bioluminescence in an isolated system. *Proc. Natn. Acad. Sci.*, US, **33,** 342.

SHARPE, A. N. & KILSBY, D. C. (1971). A rapid, inexpensive bacterial count technique using agar droplets. *J. appl. Bact.*, **34,** 435.

SHARPE, A. N. & WOODROW, M. N. (1971). Insoluble radioactive DNA complex for determination of *Staphylococcus aureus* deoxyribonuclease activity. *Analyt. Biochem.*, **41,** 430.

SHARPE, A. N., WOODROW, M. N. & JACKSON, A. K. (1970). Adenosinetriphosphate (ATP) levels in foods contaminated by bacteria. *J. appl. Bact.*, **33,** 758.

WILLIAMS, M. L. B. (1971). The limitations of the DuPont Luminescence Biometer in the microbiological analysis of food. *Can. Inst. Food Technol. J.*, **4,** 187.

Sampling of Food and Other Material for Bacteriological and Ecological Studies

BETTY C. HOBBS, W. CLIFFORD, A. C. GHOSH, R. J. GILBERT, MARGARET KENDALL, DIANE ROBERTS AND ANTONNETTE A. WIENEKE

Food Hygiene Laboratory, Central Public Health Laboratory, Colindale Avenue, London NW9 5HT, England

Introduction

The collection of adequate and suitable samples for microbiological examination is an essential part of investigations following outbreaks of food poisoning, of general surveillance studies and of food control.

Decisions are required on the choice of sample, the number of samples from a batch to give significant results, the method of sampling and the means of transport to allow little change in microbial content. Adequacy and efficiency of sampling as well as speed in delivery are as important as the methods of preparation and techniques used for examination. The results must have meaning in terms of fitness for consumption.

The well-known and worldwide microbial agents of food poisoning belong to the *Salmonella*, *Staphylococcus* and *Clostridium* groups of bacteria. *Vibrio parahaemolyticus* food poisoning is prevalent in Japan: *Bacillus cereus*, other *Bacillus* spp and group D streptococci are reported from time to time as agents of food poisoning in various countries.

The source of *Staphylococcus aureus* is usually not difficult to find; the hands of those touching warm or cold cooked foods especially meat and poultry are responsible for the majority of outbreaks of staphylococcal food poisoning. The organism may be habitually present on the hands or it may originate from the nose or from a septic lesion somewhere on the body. Sufficient evidence that the cause of an outbreak was the growth of *Staph. aureus* in food due to careless handling and poor storage is obtained from counts of *Staph. aureus* and from phage-typing and enterotoxin tests on food samples and cultures.

The sequence of events leading to *Clostridium welchii* food poisoning is also simple to follow. A history of pre-cooked meat or poultry dishes with some hours of storage at ambient or warmer temperatures before eating, in association with typical incubation periods and clinical symptoms should lead to the incrimination of the food vehicle. The isolation from one food sample of *Cl. welchii* in large numbers and of serotypes identical with those isolated from faecal samples of victims is accepted as proof of *Cl. welchii* food poisoning. The source of the spores and vegetative cells of this organism is legion and include the human and animal intestine and many foodstuffs. Control measures are centred around speed of cooling and efficient cold storage if foods are not eaten immediately after cooking. Proper cooking plays a part also but even so viable spores may remain in the food. Cold storage of cooked meats is an important preventive measure against all types of bacterial food poisoning.

The epidemiology of salmonella infection is more difficult to follow, and relatively few family outbreaks or sporadic cases, which may have a common origin, are traced either to the food vehicle or to the source of the organism (Vernon, 1970); the food source of general outbreaks is sometimes traced back to raw meat, poultry and raw milk. It is important to seek out the roots and paths of spread, because, contrary to previous conceptions it seems that salmonellae rarely reach food directly from the human excreter and regulations which only seek to control salmonellae by banning the excreter from handling food are unlikely to be successful in prevention.

Published reports and current investigations lead more and more to the conclusion that salmonellae (with the exception of *Salmonella typhi*) originate from animal excreters and reach the human population in the protein foods provided for human consumption by animals and birds. The association is rarely direct so that even when foods are adequately thawed and cooked a few surviving organisms, or those picked up from the contaminated environment including hands and utensils, may grow in the meat when stored at ambient temperatures. Thus it is necessary to work through the ramifications of contamination after cooking, and to teach that there are dangers of spread of infection from raw to cooked foods by means of hands, surfaces, equipment and utensils. Attention has now been focused on the environment, not only of the food establishment, but further back to the places where animals are reared and slaughtered and where consumable parts are processed.

To avoid exposing the housewife and her family and the clients of large-scale eating establishments to the spread of salmonellae from raw foods, sampling plans and specifications should aim to accept or reject consignments of food on the basis of criteria such as the proportion of samples containing salmonellae.

General plate counts and indicator organisms also play a part. The general counts should be low so that when there is a special risk, for example from *Staph. aureus* on seafoods peeled by hand after cooking, these organisms will be low in number also.

Calculations leading to an assessment of the number of food samples to be examined and the number of failures allowed or passes required, or the description of results required for passes and failures of any product will depend on a number of factors:

(a) Previous experience of the hazards associated with the particular product.

(b) Up-to-date knowledge of bacteriological data from many hundreds or thousands of samples examined from many sources in various countries to find a practicable level for specifications.

(c) Industrial, political and economic consequences of wholesale condemnation.

(d) The practicability of obtaining a pathogen-free product either by (i) pathogen-free livestocks, or (ii) the introduction of a terminal process to eliminate salmonellae, for example, by sterilization or pasteurization of the finished product before distribution.

(e) Provision of laboratory facilities for sampling and examination by recommended methods.

(f) Application of agreed sampling schemes and methods of examination for both home produced and imported foods, and systems of acceptance and rejection agreed between all health authorities, with constant review.

The pictures and charts which follow aim to show the establishments which merit particular attention in the search for sources of salmonellae and other food poisoning organisms and the sites which should be examined when tracing the spread from origin through to shop and kitchen. The numbers of samples and quantities suggested are based partly on practical experience over 25 years and partly on recommendations from the International Commission on Microbiological Specifications for Foods whose book on sampling is being compiled.

Methods of Sampling

Surface and equipment

Swabs

Alginate wool swabs (Higgins, 1950) on wooden sticks are packed in suitable containers and sterilized by autoclaving. Two swabs are used for each surface area, for which a template may be required, the first moistened with quarter-strength Ringer's solution or other diluent and the second used dry to mop up moisture. Both are broken off into the same bottle containing 9 ml of Ringer or other diluent, to which 1 ml of 10% sodium hexametaphosphate is added and the suspension shaken well to dissolve the alginate wool.

Agar contact

The method may be used to indicate whether surfaces are clean or dirty, rather than as a means to obtain "direct" counts; agar sausages (ten Cate, 1963, 1965) need a palette knife or other fine blade to cut slices for incubation in sterile Petri dishes. Special Petri dishes filled with agar medium for application directly to surfaces may be used also. These methods provide good visual aids.

Scrapings

Samples (1–5 g) are taken from wooden surfaces or carcasses, with a sterile knife or other sharp sterile instrument, into small sterile screw capped glass jars or polythene bags for addition to diluent or enrichment liquid.

Articles

Whole cloths, sponges and brushes, or cut portions, are placed in sterile screw capped jars or polythene bags, for addition to enrichment liquid.

Faecal samples

Human

Small quantities of faeces are taken with sterile plastic or wooden spoons into sterile screw capped glass jars. Rectal swabs are not recommended.

Animal (pen floor and litter)

Samples (5–10 g) are taken with sterile wooden applicators or spoons into sterile screw capped glass jars or waxed cartons. Alternatively a polythene bag may be used like a glove for collecting samples; the bag is then pulled inside out for the retention of the sample. The bag should be securely fastened and placed inside a second bag also securely fastened. Cotton-wool swabs may be used to obtain small quantities of faeces and caecal contents in the slaughterhouse.

Sewage

Samples (100 ml) are ladled into sterile screw capped glass jars or bottles.

FIG. 1. Farm: investigation and surveillance.

	Sample	Quantity and number	Examination
Sewer swab	Drains, gully traps	1/drain	*Salmonella*
Pen	Floor droppings, bedding	5/pen, 10g quantities bulked	*Salmonella*
Water	Mains, troughs	1000 ml	*Salmonella*
Feed	Pen, stores (Compound ingredients)	10 × 100 g	*Salmonella*
Animal	Droppings, cloacal swabs	5/pen	*Salmonella*
Milk and cream		100 g or ml	*Salmonella* *Brucella* *Staph. aureus* Other organisms Methylene blue reduction Colony counts
Cheese		100 g	*Staph. aureus* *Salmonella* Colony, coliform, *E. coli* counts Other organisms

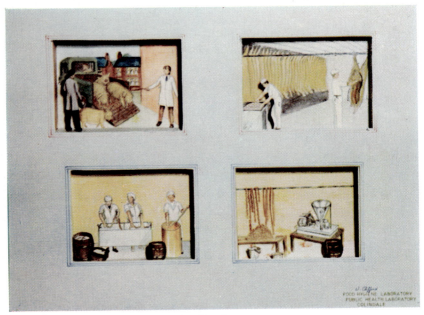

Fig. 2.

Fig. 3.

Fig. 2. (See facing page.) Slaughterhouse and meat factory: investigation and surveillance.

	Sample	Quantity and number	Examination
Sewer swab	Drains, gully traps	1–2/drain or trap	*Salmonella*
Lairage	Faeces from pen floor	Composite 5 × 20 g portions	*Salmonella*
Water	Mains, troughs from lairage, gut washing tank	1000 ml	*Salmonella* Colony, coliform, *E. coli* counts
Carcass	Composite caecal and rectal contents	10–15 g/animal	*Salmonella*
	Mesenteric lymph nodes	1–2/animal	*Salmonella*
Equipment and surfaces	Benches, mincers, slicing machines, floor	2 swabs or 1–5 g scraping/surface or article	*Salmonella* *Staph. aureus* Colony, coliform, *E. coli* counts
Product	Meat, sausages, offal Poultry	5–10 × 100 g 5–10 carcasses	*Salmonella* *Staph. aureus* *Cl. welchii* Colony, coliform, *E. coli* counts

Fig. 3. (See facing page.) Imported food: investigation and surveillance.

	Sample	Quantity and number	Examination
Frozen bulked egg	Whole, albumen	10 × 100 g/batch or shipment	—amylase, or *Salmonella*
Seafood	Frozen, cooked and raw prawns, shrimps, lobsters, crabs	10 packets or 10 × 100 g/batch or shipment	*Salmonella* *Staph. aureus* Colony, coliform, *E. coli* counts
	Fresh oysters, mussels, winkles and scallops	5–10	*Salmonella* *Staph. aureus* *Vibro parahaemolyticus* Colony, Coliform *E. coli* counts
Raw meat and poultry	Carcass and boned out beef, veal, lamb, pork, rabbit, horse, kangaroo, poultry, frogs' legs	5–10 × 100 g/batch or shipment	*Salmonella*
Cans	Shelf stable cured e.g. corned beef, luncheon meat Shelf stable uncured e.g. stewed steak, soup, vegetables Perishable cured e.g. pork, ham and shoulders	5–10 unopened cans of same code including those with defects	*Clostridium* *Bacillus* *Salmonella* *Staph. aureus* Colony, coliform, *E. coli* counts
Dehydrated products	Desiccated coconut, whole egg, yolk and white, milk, soup, seafoods, noodles, meats, spices, animal feeds	5–10 × 100 g/batch or shipment	*Salmonella* *Staph. aureus* *Cl. welchii* Colony, coliform, *E. coli* counts
Cheese	Cheddar	100 g/vat/day's production	*Staph. aureus* Colony, coliform, *E. coli* counts

Fig. 4.

Fig. 5.

FIG. 4. (See facing page.) Shop: inspection and surveillance.

	Sample	Quantity and number	Examination
Product	Raw meat, sausages, poultry; sliced cooked meat, poultry, cooked and cured sausages, pies; cream, other dairy products	100 g (inspection) 100 × 100 g over a period (surveillance)	*Salmonella* *Staph. aureus* *Cl. welchii* Colony, coliform, *E. coli* counts
Surface	Chopping block, table, counter, display cabinet, scales	1–5 g scrapings or 2 swabs/surface or article	*Salmonella* *Staph. aureus* *Cl. welchii* Colony, coliform, *E. coli* counts
Equipment	Slicing machine, can openers, utensils (knives, trays), cloths	Swabs 2/article or whole cloth Agar contact	*Salmonella* *Staph. aureus* *Cl. welchii* Colony, coliform, *E. coli* counts

FIG. 5. (See facing page.) Kitchen: investigation of outbreaks.

	Sample	Quantity and number	Examination
Food	Left-over portions, ingredients; raw, cooked, canned, packaged for humans and pets	100 g	*Salmonella* *Staph. aureus* *Clostridium* *Vibrio sp.* *Bacillus cereus* Colony, coliform, *E. coli* counts
Surface	Chopping board, table tops, shelves, cabinets, trays, dust	1–5 g scrapings, 2 swabs/surface or article	*Salmonella* *Staph. aureus* *Clostridium* Other organisms Colony, coliform, *E. coli* counts
Equipment	Refrigerator, utensils, cloths, garbage cans, can openers	2 swabs, whole article or rinsings	*Salmonella* *Staph. aureus* *Clostridium* Other organisms Colony, coliform, *E. coli* counts
Water	Supply and in use	100 ml	*Salmonella* Colony, coliform, *E. coli* counts
Person	Stool	3 × 5 g	*Salmonella* *Staph. aureus* *Clostridium* *E. coli* *Shigella* *Vibrio* sp. *Bacillus cereus* Other organisms
	Swabs, nasal, hand, throat	2 of each	*Staph. aureus* *Streptococci*

Sewers and effluents

Swabs are made up of three or more pads of gauze. Pieces of gauze, approximately 20×15 cm, are folded in four lengthways and doubled across to make a pad approximately 10×4 cm. The pads are tied together with string long enough to allow for suspension into the flow. They are sterilized by autoclaving in screw capped glass jars. After suspension for 24–48 h individual pads from the swab can be used for addition to different liquid enrichment media. The method is based on that of Moore (1948).

Water

Treated piped main supplies (chlorinated)

The outside and inside surfaces of taps are thoroughly cleaned, and the water allowed to flow for 2–3 min. The tap is sterilized by means of a blow lamp, gas torch, spirit lamp or by ignited spirit on cotton wool. The water is run to waste for about 30 sec to cool the tap before filling a sterile glass stoppered bottle containing a crystal of sodium thiosulphate; *c.* 100 ml of water are required for the most probable number (MPN) coliform and *Escherichia coli* counts (Public Health Laboratory Service, 1969).

Untreated wells and tanks

A sterile glass stoppered bottle held by the base is plunged below the surface of the water, the mouth should be directed towards the current or pushed forward horizontally until filled. Deep wells or tanks may be sampled by suspending the bottle into the water by means of a length of string sterilized with the bottle. For deep wells extra string should be provided. Approximately 100 ml of water are required for the MPN coliform and *E. coli* counts and a litre or more for enrichment procedures (Public Health Laboratory Service, 1969).

Polluted sources (troughs for lairages, gut washing and streams)

Samples may be taken in the manner described for wells and tanks or sterilized stainless steel jugs or ladles may be used to transfer the water sample to a large sterile bottle. One litre or more is required to allow for dilution with concentrated enrichment media or for the filtration technique.

Wash and rinse water from washing up sinks and tanks

A sterile glass stoppered bottle (with sodium thiosulphate) is lowered into the water by hand or by string and held until filled.

Food

In general food samples are taken with suitable implements into sterile screw capped glass jars of convenient size, waxed cartons or two polythene bags, an inner and outer one for strength, both carefully sealed. Approximately 100 g of sample is required.

The equipment required for sampling a variety of foods includes knives, scissors, secateurs, forceps, spoons, can openers (variety of types), spare keys for cans, core borers, hammer (for hard frozen foods), pincers, saw, brace and bit, chopper or meat cleaver, tongue depressors, trays, blenders (Atomix) with lids, pestles and mortars, screw capped glass jars of various sizes and polythene bags of various sizes both for samples for examination and for disposal of left-over samples. Many of these articles require to be cleaned and sterilized frequently so that good quality material and simple design is essential. They should be easy to clean and durable to repeated sterilization and if they have handles these should be constructed of materials other than wood. The materials used for cleaning and sterilization include detergents (compatible with the water), boiling water, hypochlorite and spirit (at least 70%); a spirit lamp may be necessary.

Frozen foods

Samples must be kept frozen during transport to the laboratory. Some foods, such as egg products and chickens, must be thawed before examination, others, such as frozen cooked seafoods, may be kept frozen while the sample is prepared for examination. Thawed foods should not be refrozen, except in an emergency.

Bulked whole egg in cans

The surfaces of the cans must be cleaned and sterilized with spirit before opening. Samples are drilled out while frozen with a brace and bit or electric drill or are lifted out when slightly soft with a cheese trier; a number of bits or triers will be required when many cans are sampled, they should be boiled in water between samples. Frozen flakes of egg are transferred to sterile screw capped jars, already containing enrichment media, by means of sterile spoons.

Raw or cooked seafoods

Whole packs should be sampled; portions of large packs may be chopped off into double polythene bags. In the laboratory thawed samples, or frozen samples crushed with a hammer, are homogenized in Atomix blenders and distributed into sterile screw capped jars for counts and liquid enrichment.

Raw meat from boneless blocks

At least 100 g of meat is severed by means of a cleaver or saw and allowed to drop into a polythene bag which is knotted before insertion into a second bag, also closed carefully. Instruments are cleaned and sterilized between samples. The samples, thawed or frozen, are removed from the bags and placed on enamel or stainless steel trays. They are then held with forceps and cut with knives or scissors and transferred to jars containing enrichment media.

Raw meat from carcasses

Portions of meat from the neck, anal and skirt regions are cut or chopped into polythene bags which are knotted before insertion into second polythene bags which are also closed carefully. The thawed or frozen samples on trays are cut straight into jars containing enrichment media. Knives and other equipment are cleaned and sterilized between samples.

Raw poultry

Individual polythene bags may be used both for wrapped and unwrapped carcasses or portions. On arrival in the laboratory the samples are removed from the outer bag, if present, placed on trays and allowed to thaw in the refrigerator. When thawed any remaining wrapping or bag is removed and the samples are examined by one of two methods.

Technique A. Each chicken is divided into four samples, and each sample is divided between two liquid enrichment media. The samples are edible offal or giblets chopped into small pieces, portions of skin removed from the wing areas and two sections of the backbone and ribs (where blood and debris collect inside the chicken during thawing). To obtain the last samples the chicken is cut along the sternum, the carcass opened out, and the ribs cut each side of the backbone. The length of backbone with attached pieces of rib is removed and cut into two.

Technique B (adapted from that of Surkiewicz, Johnston, Moran and Krumm, 1969). Giblets are removed, chopped and distributed into two liquid enrichment media. Three hundred ml of 1% lactose broth are added to the chicken in a plastic bag. The bag is closed and shaken thoroughly for about 30 sec. The rinse is divided between three jars: an equal volume of double strength liquid enrichment medium is added to two of the jars. All jars are then incubated.

Chilled, dehydrated, canned and other foods

The foods should not be frozen pending examination but must be kept chilled or at least cool in insulated containers during transport. The transit time for samples should be as short as possible.

Chilled meat and poultry

Samples are examined in the same way as the frozen products but as soon as possible after arrival in the laboratory. Scrapings and swabs may be taken from surfaces. Composite caecal and rectal contents (5–10 g) are taken into sterile screw capped glass jars or waxed cartons with sterile wooden applicators.

Powdered and flaked foods (including animal feeds)

Samples are taken with sterile spoons into sterile screw capped glass jars. For some epidemiological purposes, not associated with inspection, polythene bags or waxed cartons with screw caps may be used. It should be noted that contaminated powdered foods packed in polythene bags and waxed cartons may leak out so that those unpacking and testing the material could be exposed to infection and admixture of samples during transport could occur. Samples should be added to enrichment or pre-enrichment media in jars.

Cheese

Core samples from rounds or blocks of cheese are taken by means of triers or core borers. In the laboratory they are ground in a pestle and mortar with a warm solution of sodium citrate for staphylococcal counts and other tests if required. Recently satisfactory results have also been obtained with the Stomacher (see p. 206).

Cans

Sealed cans should be examined in a fume cupboard or safety cabinet. Normal cans should be tested for vacuum pressure by means of special apparatus. If the can is blown, two clean polythene bags should be placed over the cleaned top and the can opener inserted through small holes made in each polythene bag. Sterile cork borers are used for taking core samples, sterile tongue depressors or knives are used for surface scrapings and cotton-wool swabs for seam samples. Cans with liquid contents can be sampled with a sterile spoon. All samples are taken into sterile screw capped glass jars for examination.

Sampling of Bacterial Colonies in the Investigation of Food Poisoning Outbreaks and in Epidemiological Studies

In the bacteriological investigations of food poisoning outbreaks the isolation of one serotype or phage type of a pathogen is usually accepted as evidence of the cause, although in *Salmonella* outbreaks for example, multiple serotypes may be involved (Hormaeche, Surraco, Peluffo and Aleppo, 1943; Juenker, 1957; Taylor, 1960). It is impracticable to identify and type by sera or phages every colony of *Salmonella*, *Cl. welchii* or *Staph. aureus* on a plate of culture medium. Conversely, when a single colony per plate is examined evidence related to the source may be missed.

The detection of multiple serotypes or phage types is also important in epidemiological studies. In 1959 for example, P.M. Lynne (pers. comm.) isolated 19 serotypes of *Salmonella* from a sample of imported dried egg albumen by serotyping a large number of colonies from various plates of selective media and Harvey and Price (1962) isolated 17 serotypes of *Salmonella* from one sample of Indian crushed bone using a serological technique.

Examples are given where the examination of a number of similar colonies on a single plate has revealed different types. Care and patience may be needed but the results can be rewarding.

Salmonella

More than 1600 serotypes are included in this genus and routine samples of meat, sausages, poultry and animal feeding stuffs frequently yield the organism. Two $\times$ 25 g aliquots of each sample are usually incubated in two different enrichment media and subcultures are made after 1 and 3 days on to 2 selective media (i.e. 8 plates of selective media per sample). If suspicious colonies are observed at least one should be picked from each plate of selective medium.

Harvey and Price (1967) described 6 techniques useful for the examination of foods, water samples and feeding stuffs containing multiple serotypes of *Salmonella*. The simplest method was to pick and serotype numerous colonies from selective agar media. Table 1 shows results obtained by this technique.

TABLE 1. *Salmonella* serotypes isolated from single plates of selective media

No.	Sample	Selective medium*	No. of colonies tested	Serotypes isolated	Comments
1	Pork sausage	BGA	10	5 — unnamed (4,12:d:−) 5 — anatum	
2	Pork sausage	BGA	10	9 — infantis 1 — anatum	S. anatum was not isolated from other plate cultures
3	Pork sausage	BGA	5	3 — derby 2 — panama	Colonies of S. derby and S. panama were visibly different
	Pork sausage	DCA	6	4 — derby 2 — panama	
4	Pork sausage	BGA	10	9 — infantis 1 — reading	Only 10 suspect Salmonella colonies on the plate
5	Pork sausage	BGA	10	6 — infantis 4 — unnamed (4,12:d:−)	
6	Chicken	DCA	18	16 — typhimurium 2 — livingstone	The first 2 colonies picked were S. livingstone—the next 16 were all S. typhimurium
7	Chicken	DCA	9	8 — bredeney 1 — livingstone	
	Chicken	BSA	5	4 — bredeney 1 — livingstone	
8	Pig faeces	BSA	c.10	?— brandenburg ?— typhimurium	
9	Pig faeces	BSA	c.10	?— brandenburg ?— typhimurium 13 — oranienburg 11 — anatum	
10	Crushed bone†	BSA	50	10 — karachi 9 — kirkee 5 — jodpur 1 — bronx 1 — richmond	Had only 10 colonies been picked only 4 serotypes would have been isolated

* BGA, Brilliant Green Agar; DCA, Deoxycholate Citrate Sucrose Agar, and BSA, Bismuth Sulphite Agar.
† Table 1 from Harvey and Price (1967).

Clostridium welchii

Cultures of *Cl. welchii* (toxicological type A) from most food poisoning outbreaks in the UK are serotyped with two sets of antisera, one for heat-
resistant (usually non-haemolytic) strains (Hobbs types 1–24) and the other
for the so-called heat sensitive (usually haemolytic) strains (types i–xviii).
About 65% of cultures isolated from outbreaks can be typed with the
present range of antisera.

In many outbreaks of *Cl. welchii* food poisoning a single serotype is isolated from food samples and from faecal specimens from patients. In
others, mixtures of both heat-resistant and heat-sensitive types may be
isolated and assumed to be implicated (Sutton and Hobbs, 1968). Outbreak
A (Table 2) provides good evidence that multiple serotypes of heat-
resistant *Cl. welchii* were involved. Two types, 5 and 7, were present in the

TABLE 2. Serological types of *Clostridium welchii* from two outbreaks of
food poisoning

Outbreak	Sample no.	Sample	Serological type of *Cl. welchii* isolated	
			Heat-resistant	Heat-sensitive
	1	Faeces	5 and 6	—
	2	Faeces	5 and 7	—
	3	Faeces	5 and 7	—
	4	Faeces	5 and 7	—
	5	Faeces	5 and 7	—
A	6	Faeces	5, 6 and 7	—
	7	Faeces	6 and 7	—
	8	Faeces	6 and 19	—
	9	Faeces	6, 7 and 19	—
	10	Faeces	7 and 19	—
	11	Boiled beef	5 and 7	—
	1 to 12	Faeces	—	ii
	13	Faeces	4	ii
	14	Faeces	5	ii
	15	Faeces	—	ii and xiv
B	16	Stewed beef*	—	iii
	Re-examination of stewed beef—12 separate colonies submitted for examination			
	7 colonies		—	ii
	3 colonies		—	iii
	1 colony		—	xiv
	1 colony		—	xv

* One colony only, from blood agar was submitted for examination.

food and in at least 6 faecal specimens; two other types, 6 and 19, were also isolated from some of the faecal specimens. The results from outbreak B show that by picking one colony only from the food culture, in this instance type iii, the causal agent, shown by faecal isolations to be type ii was missed. Due to foresight the laboratory investigating the outbreak had kept the samples, so fresh cultures from the food were made. Twelve separate colonies were submitted for typing. Of these, 7 were found to be type ii, 2 were type iii, 1 was type xiv and 1 type xv.

It is recommended that at least 5 colonies of *Cl. welchii* be picked from primary cultures grown from the food and sent as separate strains for typing.

Staphylococcus aureus

Cultures of *Staph. aureus* from food poisoning outbreaks can be tested for enterotoxin production and if the relevant foods are still available, tests for enterotoxin in the food can be made also.

During the examination of routine food samples colonial variation in pigmentation and haemolysis on horse blood agar have been observed. Tests for phage-type and enterotoxin production showed that these foods contained more than one strain of *Staph. aureus*. Thus colonial variation should not be ignored. In outbreak A, *Staph. aureus* was isolated from a sample of ham and from two vomit and two faecal specimens. Four of the strains formed non-haemolytic colonies on blood agar, but the fifth (from ham) produced two types of colony, non-haemolytic and a small number which were α-β haemolytic. Of 11 non-haemolytic colonies tested all produced enterotoxin A (10–15 μg/ml) when grown in separate sac cultures at 35° for 1 day on a rotatory shaker. All the colonies were lysed by various phages of group III including phage 85; none were lysed by phage 81 (miscellaneous group). Six α-β haemolytic colonies also produced enterotoxin A but in smaller amounts (2·5–3·7 μg/ml). These colonies were also lysed by various phages of group III but not phage 85; they were, however, lysed by phage 81.

In outbreak B, *Staph. aureus* was isolated from a sample of ham and from one faecal specimen. The culture from faeces produced white colonies on blood agar, whereas the culture from ham produced not only white colonies but also a small number of yellow colonies. The phage typing results indicated that the colonies were indistinguishable. However, each of 10 white colonies from the ham culture produced enterotoxins A (2·5–5 μg/ml) and C (200–400 μg/ml), whereas 10 of the yellow colonies produced enterotoxin C (200–400 μg/ml) only.

Comments

There are two general precepts associated with the sampling of food. The first, that samples must be accompanied by adequate data on their history; Tables 3 and 4 give examples of information required with foods from outbreaks and from surveillance studies. A fair assessment of the significance of the results of examination can only be made when such information is available.

The second, that efforts should be made to find the source of contaminating organisms which have or which could cause food poisoning.

TABLE 3. Questionnaire for food poisoning outbreaks

Area and place of outbreak . . . .
Date and time suspected meal eaten . . . .
Number affected . . . . Number at risk . . . .
Incubation period . . . .
Symptoms . . . .
Occupation and approx. age group . . . .
Details of suspected meal . . . .
Foods common to all affected persons . . . .
Foods not eaten by unaffected persons . . . .
Number of meal sittings and times . . . .
Methods of cooking (particularly meat and poultry) . . . .
Time and temperature of storage after cooking . . . .
Foods sampled . . . . Place of sampling . . . .
Transport time and temperature . . . .
General notes on facilities and environment . . . .

TABLE 4. Basic information required for surveillance studies on imported or home produced food

Authority . . . .
Date of arrival . . . .
Type of sample . . . .
Number of sample . . . .
Brand, code, weight . . . .
Size of consignment . . . .
Country of origin . . . .
Manufacturer and date of manufacture . . . .
Ship/retailer . . . .
Port/wharf . . . .
Storage since manufacture and during transport . . . .
    (Time and temperature)
Transport to laboratory . . . .
    (Time and temperature)

For example, salmonellae in animal feeding stuffs work their way through poultry, pigs and other animals to carcass meat and other raw products. Raw foods such as boned out and comminuted meat (mince and sausages) and poultry are those most likely to introduce salmonellae into kitchens, but the source of the organism may be a long way back in the chain. It should be noted that sporadic cases as well as general outbreaks of food poisoning may have a common food vehicle. The relevant serotypes of salmonellae are more likely to be isolated from the raw materials than cooked food, unless the immediate vehicle of infection is available; often it is eaten or destroyed. Similarly, in surveillance studies the examination of large numbers of cooked foods for salmonellae is usually unrewarding.

Food samples which are too small in size, too few in number, and even inappropriate to the ecology of the organism concerned in an outbreak are valueless. Sampling wrong products is wasteful of time; days, weeks and months may be spent in fruitless examinations. Therefore, the phrase "look before you leap" should be applied to sampling, because first hand knowledge is now available on sources, routes of spread and behaviour patterns of the common microbial agents of food poisoning. Persistent education by teaching and writing is required to highlight the points where investigation, examination and control are appropriate.

After sampling comes the choice of laboratory methods of examination; there are many books and manuals on this subject. An assessment of numbers and quantities of samples required in relation to statistically sound formulae for acceptance and rejection of different foods is under international appraisal.

References

TEN CATE, L. (1963). Eine einfache und schnelle bakteriologische Betriebskontrolle in Fleisch verarbeitenden Betrieben mittels Agar-Wursten in Rilsan-Kunstdarm. *Fleischwirtschaft*, **15**, 483.

TEN CATE, L. (1965). A note on a simple method of bacteriological sampling by means of agar sausages. *J. appl. bact.*, **28**, 221.

HARVEY, R. W. S. & PRICE, T. H. (1962). Salmonella serotypes and arizona paracolons isolated from Indian crushed bone. *Mon. Bull. Minist. Hlth*, **21**, 54.

HARVEY, R. W. S. & PRICE, T. H. (1967). The examination of samples infected with multiple salmonella serotypes. *J. Hyg., Camb.*, **65**, 423.

HIGGINS, M. (1950). A comparison of the recovery rate of organisms from cotton-wool and calcium alginate swabs. *Mon. Bull. Minist. Hlth*, **9**, 50.

HORMAECHE, E., SURRACO, N. L., PELUFFO, C. A. & ALEPPO, P. L. (1943). Causes of infantile summer diarrhoea. *Am. J. Dis. Child.*, **66**, 539.

JUENKER, A. P. (1957). Infections with multiple types of salmonellae. *Am. J. clin. Path.*, **27**, 646.

MOORE, B. (1948). The detection of paratyphoid carriers in towns by means of sewage examination. *Mon. Bull. Minist. Hlth*, **7**, 241.

PUBLIC HEALTH LABORATORY SERVICE (1969). The bacteriological examination of water supplies. 4th Ed. *Rep. publ. Hlth med.* Subj No. 71. London: HMSO.

SURKIEWICZ, B. F., JOHNSTON, R. W., MORAN, A. B. & KRUMM, G. W. (1969). A bacteriological survey of chicken eviscerating plants. *Fd Technol., Champaign,* **23,** 80.

SUTTON, R. G. A. & HOBBS, B. C. (1968). Food poisoning caused by heat-sensitive *Clostridium welchii.* A report of five recent outbreaks. *J. Hyg., Camb.,* **66,** 135.

TAYLOR, J. (1960). The diarrhoeal diseases in England and Wales. *Bull. Wld Hlth Org.,* **23,** 763.

VERNON, E. (1970). Food poisoning and salmonella infections in England and Wales. *Publ. Hlth, Lond.,* **84,** 239.

Author Index

Numbers in italics are pages on which references are listed at the end of the paper.

Subject Index